SpringerBriefs in Ecology

For further volumes:
http://www.springer.com/series/10157

Clara B. Jones

The Evolution of Mammalian Sociality in an Ecological Perspective

 Springer

Clara B. Jones
Mammals and Phenogroups (MaPs)
Asheville
North Carolina
USA

ISSN 2192-4759   ISSN 2192-4767 (electronic)
ISBN 978-3-319-03930-5   ISBN 978-3-319-03931-2 (eBook)
DOI 10.1007/978-3-319-03931-2
Springer Cham Heidelberg New York Dordrecht London

Library of Congress Control Number: 2013958320

© Clara B. Jones 2014
This work is subject to copyright. All rights are reserved by the Publisher, whether the whole or part
of the material is concerned, specifically the rights of translation, reprinting, reuse of illustrations,
recitation, broadcasting, reproduction on microfilms or in any other physical way, and transmission or
information storage and retrieval, electronic adaptation, computer software, or by similar or dissimilar
methodology now known or hereafter developed. Exempted from this legal reservation are brief excerpts
in connection with reviews or scholarly analysis or material supplied specifically for the purpose of
being entered and executed on a computer system, for exclusive use by the purchaser of the work.
Duplication of this publication or parts thereof is permitted only under the provisions of the Copyright
Law of the Publisher's location, in its current version, and permission for use must always be obtained
from Springer. Permissions for use may be obtained through RightsLink at the Copyright Clearance
Center. Violations are liable to prosecution under the respective Copyright Law.
The use of general descriptive names, registered names, trademarks, service marks, etc. in this publication
does not imply, even in the absence of a specific statement, that such names are exempt from the relevant
protective laws and regulations and therefore free for general use.
While the advice and information in this book are believed to be true and accurate at the date of
publication, neither the authors nor the editors nor the publisher can accept any legal responsibility for
any errors or omissions that may be made. The publisher makes no warranty, express or implied, with
respect to the material contained herein.

Printed on acid-free paper

Springer is part of Springer Science+Business Media (www.springer.com)

Preface

This SpringerBrief concerns the condition-dependent effects of intraspecific competition on a mammal's reproductive rate, as well as actions available to an individual (hereafter "type" [genotype and phenotype]) in the face of competition. Group-formation, group-maintenance, and sociality may be favored by selection when genes of social types increase relative to non-social types, beyond some threshold level in a population. When "social" and "sociality" are defined as *responses facilitating the reproduction of one or more conspecific*, sociality is not a conspicuous feature of Class, Mammalia. Facilitation appears in response to density-dependent conditions characterized by "thermal stress" (stimuli negatively impacting reproductive rate) when niche spaces ("thermal zones") of (conspecific) types (genotypes and phenotypes: individuals, organisms) overlap relative to variations in resource dispersion (distribution, abundance, and/or quality of food, mates, and/or space). The previous conditions constitute intermediate or high levels of competition for limiting resources, resources influencing the reproductive rates of types. Group-formation and group-maintenance are necessary, but not sufficient, precursors to the evolution of sociality, and recent treatments show that coexistence of different types is best studied using a "trait-based approach". In theory, a type can be decomposed into a set of expressible traits with varying values dependent upon condition. Social traits expressed by types include "alloparenting", cooperation, reciprocity, and altruism, unambiguous and measurable features of phenotypes permitting independent quantitative analyses within and between populations and species.

Mammals were preadapted for solitary living during the Triassic when mammal-like reptiles escaped reptilian competitors by adopting nocturnal habits. Consistent with the ancestral patterns of extant mammals and reptiles, the former are, primarily, nocturnal and solitary, the latter, primarily, diurnal. In general, extant adult male mammals are intolerant of other males and of the young, while adult females, unless signaling sexual receptivity, are intolerant of conspecifics other than their offspring and of males. The whole-organism phenotype of one mammal is exposed to abiotic (soil, climate) and biotic (plants, predators) environments that may or may not be correlated across space and time. Ultimately, selection acts on genetically-correlated phenotypic traits and, from a genotype's perspective, copies of alleles and their associated traits may be carried or expressed throughout a population and, sometimes, a region ("metapopulation"). For a given genotype, phenotypes will vary spatiotemporally within and between individuals bearing a trait or traits. Mammals found in aggregations (temporary assemblages of one or more than one species), or spatiotemporally recurrent groups, may or may not occur in proximity to individuals bearing the same genotype because dispersing types may travel near ("viscosity") or far from their natal groups. A type will be designed to do the best it can do to maximize its relative fitness (growth rate) in a population, even though niche spaces of similar types will overlap, yielding competition for limiting resources (e.g., food, mates, space). Thus, interests of types may not coincide, particularly kin whose niche spaces are bound to overlap, and responses influencing reproductive rates of similar or different types may have beneficial or deleterious effects on per capita rates of population growth.

Expanding other treatments, intensities of within- and among-species competition in local ("patch") and regional (population, metapopulation) regimes are expected to determine benefits and costs to each mammal's current and future reproduction via condition-dependent interactions between genotype and environment ("reaction norms"), including interactions with other members of an aggregation or integrated group. Throughout the present review, mammals are assumed to reside in a competitive context, within the individual mammal's group, between groups, and within populations, communities and ecosystems, and interactions between or among conspecifics may be categorized as facilitation, tolerance, or inhibition. The topic of the present synthesis is mammalian social evolution, and, throughout the text, "facilitation" is employed generically to mean facilitation of a type's relative reproductive interests via the facilitation of another type's reproductive interests, usually another group member and often a relative.

Evolutionary transitions to sociality within and between mammalian taxa are central to a scientific understanding of sociality as a phenomenon, since the Class constitutes the most ecologically dominant terrestrial vertebrate fauna, including grades of population structure from "solitary" ("sexually-segregated") to eusocial (overlapping generations, cooperative breeding, reproductive division of labor). Sterile castes have, apparently, not evolved among mammals; thus, in the present brief, cooperatively breeding and eusocial molerats are classified as "primitively eusocial".

In addition, this monograph discusses factors associated with group-formation, group-maintenance, group population structure, and, other, events and processes (e.g., physiology, behavior). Within- and between-lineages, features of prehistoric and extant social mammals, patterns and linkages are discussed as components of a possible social "tool-kit", and "top-down" (predators to nutrients), as well as, "bottom-up" (nutrients to predators), effects are assessed. The present synthesis also emphasizes outcomes of Hebbian (synaptic) "decisions" on Malthusian parameters (growth rates of populations) and their consequences for (shifting) mean fitnesses of populations. Ecology and evolution (EcoEvo) are connected via the organism's "norms of reaction" (genotype × environment interactions; life-history tradeoffs of reproduction, survival, and growth) exposed to selection, with the success of genotypes influenced by intensities of selection as well as neutral (e.g., mutation rates) and stochastic effects. At every turn, life history trajectories are assumed to arise from "decisions" made by types responding to competition for limiting resources constrained by Hamilton's rule (inclusive fitness operations).

This and previous projects would not have been possible without efforts, constructive criticisms, and other inputs from many individuals. I thank Janet Slobodien and her staff for facilitating the current project. My graphics assistants, Monica E. Mc-Garrity, Liz Williams, as well as Charla Schlueter, deserve mention for their technical talents and patience. I am very grateful to Kenneth D. Angielczyk, Lee Drickamer, Ted Fleming, Steven A. Frank, Tarmo Ketola, Phyllis C. Lee, Jesse Marczyk, Daniel J. Mennill, Peter Nonacs, Craig Packer, Michael Platt, Randy Thornhill, Robert L. Trivers, Gene E. Robinson, two anonymous reviewers and, especially, Andrew Bourke, for responding to my questions or commenting on one or more sections of this brief. My children, Dalton Anthony, Julie Karin, and Miguel Luke Jones provided insight, good humor, and support. This brief is dedicated to my teachers, mentors, and colleagues.

Contents

Chapter 1
Introduction: Definitions, Background

The 'norm of reaction' of a trait refers to its range of expressed phenotypes plotted as a function of changes in the environment.

Qvarnstöm (2001)

We can understand the differences in population level output as a function of differences in individual-level parameters.

Martin et al. (2013)

The specializations for a carnivorous way of life are less extreme than those demanded of plant-eaters. In consequence it is the carnivores and insectivores which give rise to new groups of vertebrates.

Kermack and Kermack (1984)

Abstract This chapter provides an overview of mammalian social evolution, including selected history and preliminary definitions. The primary emphasis is a brief discussion of "routes" to sociality, constrained by Hamilton's rule. Mammals provide a specific as well as a general model for understanding sociality because of the wide range of structures represented, from "solitary" ("sexually segregated") to "primitively eusocial" species. On the other hand, most mammals are "solitary," exhibiting sexual segregation, or polygynous (one reproductive male monopolizing more than one reproductive female), with the potential to inform social biologists about evolutionary limits to the evolution of sociality.

Keywords History · Evolutionary transitions · Social mammals · Social actions · Routes to sociality · Convergent evolution

Crook's (1964, 1965) behavioral ecology paradigm (Fig. 1.1, Chaps. 5 and 6; see also Birkhead and Monaghan 2010) and Hamilton's (1964) inclusive fitness theory (Fig. 1.2) inspired the development of systems classifying social taxa and "routes" to sociality. These projects were driven not only by knowledge of lineages, morphological design, behavior, and ecology, but also by an understanding that convergent evolution may lead different lineages to solve environmental challenges in similar ways (Weinreich et al. 2006; Woodard et al. 2011; Bourke 2011). Crespi (2007) promoted the view that mating systems and social systems coevolve, expanding an understanding of the relationships among reaction norms, relative reproductive

C. B. Jones, *The Evolution of Mammalian Sociality in an Ecological Perspective,* SpringerBriefs in Ecology, DOI 10.1007/978-3-319-03931-2_1, © Clara B. Jones 2014

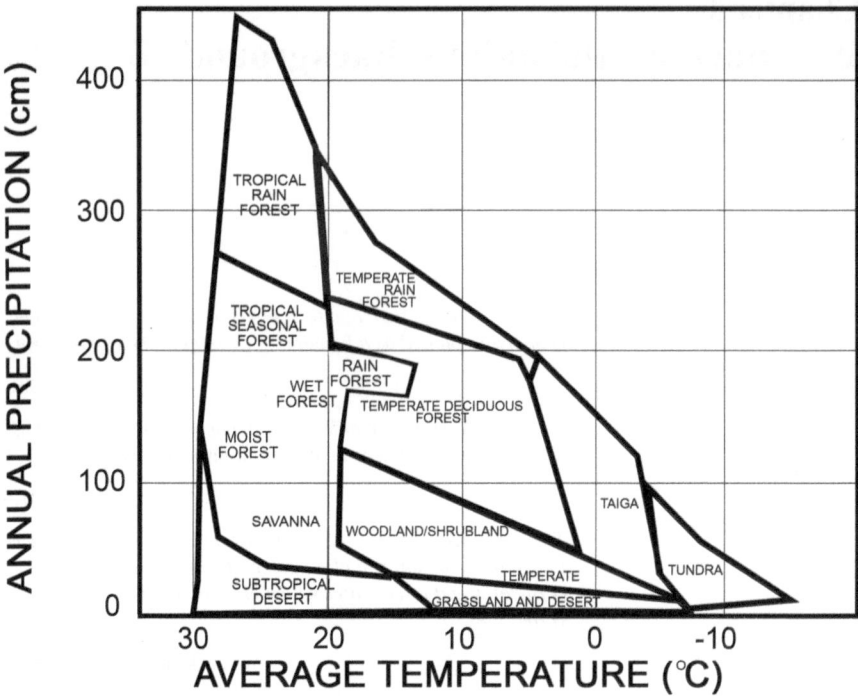

Fig. 1.1 Relative proportions of worldwide distributions of terrestrial (vegetation) ecosystems (cumulative "niche space") in which organisms survive and reproduce, integrating Whittaker's and Holdridge's biome/ecosystem maps. See text for further discussion, including Chap. 6 and 7. ©Clara B. Jones

success, and social evolution. As a result, for example, "social monogamy" is differentiated from "sexual monogamy," and, leks are, at the same time, spatiotemporally defined breeding areas as well as regimes in which a subset of conspecific males exhibit short-term, interindividual tolerance.

1.1   Different "Routes" to Sociality

According to Darwinian litany (1964), "No instinct has been produced for the exclusive good of other animals, but each animal takes advantage of the instincts of others." A contemporary rendering of Darwin's statement holds that, "Natural selection favors cooperation when genes underlying it increase in frequency compared with their non-cooperative counterparts" (Strassmann and Queller 2011). Because of the relative fitness costs to Actor imposed by altruism and spite (Fig. 1.3), demonstrating stable evolutionary scenarios of these states using quantitative approaches has proved challenging.

Fig. 1.2 Direct reproduction is defined as the biological production of one's own offspring. Hamilton's (1964; "Hamilton's rule") inclusive fitness theory is the central tenet of social evolution, a quantitative model showing that facilitating the reproduction of a relative (altruism: Fig. 1.3), indirect reproduction, may increase a type's reproductive rate. Benefit (b) to the altruist (Actor: Fig. 1.3) depends upon the number of alleles at a locus shared by the altruist and recipient (Recipient: Fig. 1.3) of the altruistic act (r), the benefit (b) to the recipient (Fig. 1.3), as well as the cost (c) to the altruist. Benefits and costs are generally measured as the number of offspring produced or as reproductive rate (e.g., inter-birth interval). Hamilton's rule, thus, states the conditions under which r, b, and c are > 0, and when altruism is expected to be favored. If expectations are not met, then some measure is inaccurate (Bourke 2011). A "decision" to perform an altruistic act reflects a strategic "decision" based upon competitive conditions in time and space, maximizing a type's reproductive rate responding to competitive regimes. A "decision" to act altruistically (or with any other response) represents one of the possible responses to manage competitive interactions with conspecifics, including recipients of the act. Presumably the traits expressed in association with any action represent the ones most likely to enhance direct or indirect reproduction of the actor. Text and Concept design © Clara B. Jones; Artwork © Liz Williams

Attempts to explain biological transitions to social states have generated several constructs. Revising and expanding Wilson's (1971) classification of social insect systems, Vehrencamp (1979) argued that the evolution of sociality in mammals is constrained where parental, almost always maternal, care terminates subsequent

	BENEFITS OR COSTS TO ACTOR	BENEFITS OR COSTS TO RECIPIENT
SELFISH	+	-
COOPERATIVE	+	+
ALTRUISTIC	-	+
SPITEFUL	-	-

Fig. 1.3 This figure presents a widely accepted schema depicting hypothetical outcomes of interactions between two conspecifics (a "dyad"), including, predicted tradeoffs (Trivers 1985). A "selfish" state is presumed to be original and fundamental since selection acts on individual genotypes and since assisting ego's own reproduction should benefit a type more than benefiting other types' reproduction. A type is always related to itself by 1.00. It follows that cooperation and altruism have evolved where a selfish strategy cannot do its best. Social biologists' assumption that sociality evolved in "poor" conditions follows from the latter inference as well as from empirical evidence (Chaps. 3–7). By definition, "cooperation" and "altruism" require Actor ("ego") to restrain some measure of its "fitness budget" in the course of maximizing reproductive rate (thus, lifetime reproductive success) by donating some measure of reproduction to a conspecific recipient. The recipient is often, but, not, necessarily, a female relative, and the currency is usually in the form of work (the transfer of energy from one system to another). Alternatively, aggressive and reproductive restraint may be imposed on one type by another type *via* mechanisms of persuasion, coercion, force, or exploitation (Table 2.1). See Chap. 2 for further discussion of these "decisions"

to weaning. In these scenarios, helping would not be favored, even among female kin. Vehrencamp (1979) goes on to point out that cooperative breeding and eusociality have evolved in some mammals, arguing that these evolutionary transitions have arisen via a "familial route" (solitary → subsocial → intermediate subsocial → eusocial) rather than a "parasocial route" (solitary → communal → quasi-social → semisocial → eusocial). In familial systems, the intermediate subsocial stage is characterized by related individuals in a group dividing labor (defense of a refugium, sharing information about food or other resources) but without reproductive division of labor featured in the final, eusocial, stage. Vehrencamp's (1979) schema has the advantage of permitting comparisons within and between taxonomic groups, highlighting consequences of transitions from solitary to advanced social architectures.

Compared to Vehrencamp's (1979, 2000) approaches, Crespi's (2007) conceptual template is a simpler one with the advantage of greater flexibility for comparative purposes, permitting highly detailed analyses on a case-by-case, condition-by-condition, basis, leading to tests of socioecological hypotheses (Chap. 6). As noted, Crespi (2007) stressed the idea that sexual systems and group population structure coevolve and that social transitions, should they arise at all, begin with ecological forces favoring group life. According to the latter author, once groups form, maternal care may be favored (obligate, in mammals), and Crespi advanced the idea that the route to and type of allomaternal care (care of offspring by group members other than the mother: mammals, Solomon and French 1997; Hrdy 1976) depend upon whether

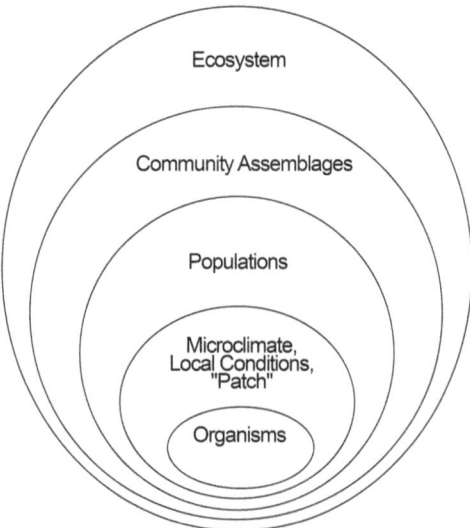

Fig. 1.4 The present brief's hierarchical design, from organism to population to community to ecosystem. "Top-down" (predators to nutrients) and "bottom-up" (nutrients to predators) effects are weighed. Challenges at the level of organisms (hereafter, type or types) involve specifying the reproductive rates of types contributing to shifting mean "demographic dynamics", "species richness gradients" and changes in diversity, as well as ecosystem processes (Martin et al. 2013; Duffy 2002). Duffy's (2002) discussion strongly suggests that bottom-up effects should be stronger in heterogeneous regimes, conditions implicated in mammalian evolution (Jones 2009; this brief, Chap. 7, Fig. 1.1). The latter effects remain to be studied in detail, particularly dynamic feedbacks "of organisms and their performance" in association with changes in population density and community stability operating in a dynamic ecosystem context (Martin et al. 2013). Suggesting that the aforementioned relationships are not straightforward, Yessoufou et al. (2013; Duffy 2002) found that, while extinction of large mammalian herbivores had "cascading effects on plant diversity", reintroductions of those taxa resulted in "mixed impacts" on "plant community structure". Indirect interactions have, also, been implicated in "cascading extinctions of carnivores" (Sanders et al. 2013). Bottom-up effects have been reported for the colonial, plains vizcacha (*Lagostomus maximus*: Villareal et al. 2008). ©Clara B. Jones (after Dalton et al. 2004)

infant care is limited to females or is shared by males. According to Crespi (2007), the evolution of biparental care facilitates the evolution of cooperative breeding and eusociality (see also, Helms Cahan et al. 2002), a testable scenario using comparative analyses of taxa displaying the latter social architectures.

A comprehensive program was advanced by Bourke (2011) whereby causes and consequences of social trajectories were assessed using inclusive fitness theory (Hamilton's rule: Fig. 1.2; also see Queller 1992), discussed in association with the evolution of transitions leading to complexity in Nature. In Bourke's (2011) system, altruism arises via one, and only one, mechanism, sharing of genes between relatives, either via a "subsocial" route (parent to offspring: vertical transmission) or a "semisocial" route (same generation, horizontal transmission between kin; see Cornwallis et al. 2009 for vertebrates). Lack of sharing of genes between unrelated conspecifics constraints social evolution to cooperation if sociality obtains at all. Thus, social evolution may be favored where types share common genes or common reproductive "fates," and altruism may evolve only when r is positive (Lehmann and Keller 2006; Bourke 2011).

The semisocial route may explain formation of social groups by unrelated females among polygynous and polygynandrous mammals. Phenotypically, mammals carrying social genes may have the ability to estimate kinship based upon an immigrating type's origin (e.g., same habitat, same population, different population), "decisions" with different likelihoods of carrying the same allele at the same locus. Mammals in heterogeneous (stochastic, fluctuating, but not necessarily unpredictable) environments should be predisposed to tolerate conspecifics from the same localities, but not necessarily from more distant regions. While fluctuating environments are correlated with the evolution of cooperation and altruism (Alexander et al. 1991; Jetz and Rubenstein 2011), they are, as well, expected to be "hotspots" for the evolution of exploitation (Galef 1991; this brief Fig. 1.2), consistent with Hamilton's rule. Similar to the social insects, the evolution of exploitation (e.g., social parasitism, cryptic "female choice," "slavemaking") among mammals should reach a pinnacle when evaluating interactions among kin and mates (Wilson 1971). A modified version of Bourke's (2011) treatment is adopted herein. I intend to show that, while the systematic study of mammalian sociality will inform scientists about promoters of and constraints on social evolution, no mammal society has reached the degree of obligatory interdependence to be considered an individual subunit in its own right (Fig. 1.4).

References

Alexander RD, Noonan KM, Crespi BJ (1991) The evolution of eusociality. In: Sherman PW, Jarvis JUM, Alexander RD (eds) The biology of the naked mole-rat. Princeton University Press, Princeton

Birkhead TR, Monaghan P (2010) Ingenious ideas: the history of behavioral ecology. In: Westneat DF, Fox CW (eds) Evolutionary behavioral ecology. Oxford University Press, Oxford

Bourke AFG (2011) Principles of social evolution. Oxford University Press, Oxford

Cornwallis CK, West SA, Griffin AS (2009) Routes to indirect fitness in cooperatively breeding vertebrates: kin discrimination and limited dispersal. J Evol Biol 22:2445–2457

Crespi BJ (2007) Comparative evolutionary ecology of social and sexual systems: waterbreathing insects come of age. In: Duffy JE, Thiel M (eds) Evolutionary ecology of social and sexual systems: crustaceans as model organisms. Oxford University Press, Oxford

Crook JH (1964) The evolution of social organization and visual communication in the weaver birds (Ploceinae). Leiden, Brill. (Behaviour Supplement X)

Crook JH (1965) The adaptive significance of avian social organization. Symp Zool Soc Lond 14:181–218

Dalton JH, Elias MI, Wandersman A (2004) Community psychology. Wadsworth, Stamford, CT

Darwin CR (1964) On the origin of species. Harvard University Press, Cambridge, MA (Facsimile of 1st Ed, 1859)

Duffy JE (2002) Biodiversity and ecosystem function: the consumer connection. Oikos 99:201–219

Galef BJ (1991) Information centers of Norwegian rats: sites for information exchange and information parasitism. Anim Behav 41:295–301

Hamilton WD (1964) The genetical theory of social behavior. J Theor Biol 7:1–52

Helms Cahan S, Blumstein DT, Sundström L, Liebig J, Griffin A (2002) Social trajectories and the evolution of social behavior. Oikos 96:206–216

Hrdy SB (1976) The care and exploitation of nonhuman primate infants by conspecifics other than the mother. Adv Stud Behav 6:101–158

Jetz W, Rubenstein DR (2011) Environmental uncertainty and the global biogeography of cooperative breeding in birds. Curr Biol 21:72–78

Jones CB (2009) The effects of heterogeneous regimes on reproductive skew in eutherian mammals. In: Hager R, Jones CB (eds) Reproductive skew in vertebrates: proximate and ultimate causes. Cambridge University Press, Cambridge

Kermack DM, Kermack KA (1984) The evolution of mammalian characters. Croom Helm, London

Lehmann L, Keller L (2006) The evolution of cooperation and altruism—a general framework and a classification of models. J Evol Biol 19:1365–1376

Martin TE, Ton R, Nikilson A (2013) Intrinsic vs. extrinsic influences on life history expression: metabolism and parentally induced temperature influences on embryo development rate. Ecol Lett 16:738–745

Queller DC (1992) A general model for kin selection. Evolution 46:376–380

Qvarnström A (2001) Context-dependent genetic benefits from mate choice. Trends Ecol Evol 16:5–7

Sanders D, Sutter L, van Veen FJF (2013) The loss of indirect interactions to cascading extinctions of carnivores. Ecol Lett 16:664–669

Solomon NG, French JA (eds) (1997) Cooperative breeding in mammals. Cambridge University Press, New York

Strassmann JE, Queller DC (2011) Evolution of cooperation and control of cheating in a social microbe. Proc Nat Acad Sci U S A 108:10855–10862

Trivers RL (1985) Social evolution. Benjamin-Cummings, Menlo Park

Vehrencamp SL (1979) The roles of individual, kin, and group selection in the evolution of sociality. In: Marler P, Vandenbergh JG (eds) Handbook of behavioral neurobiology: social behavior and communication, Vol. 1. Plenum, New York

Vehrencamp SL (2000) Evolutionary routes to joint-female nesting in birds. Behav Ecol 11:334–344

Villareal D, Clark KL, Branch LC, Hierro JL, Machicote M (2008) Alteration of ecosystem structure by a burrowing herbivore, the plains vizchaca (*Lagostomus maximus*). J Mammal 89:700—711

Weinreich DM, Delaney NF, DePristo MA, Hartl DL (2006) Darwinian evolution can follow only very few mutational paths to fitter proteins. Science 312:111–114

Wilson EO (1971) The insect societies. Belknap, Cambridge

Woodard SH, Fischman BJ, Venkat A, Hudson ME, Varala K, Cameron SA, Clark AG, Robinson GE (2011) Genes involved in convergent evolution of eusociality in bees. Proc Nat Acad Sci U S A 108:7472–7477

Yessoufou K, Davies TJ, Maurin O, Kuzmina M, Schaefer H, van der Bank M, Savolainen V (2013) Large herbivores favour species diversity but have mixed impacts on phylogenetic community structure in an African savanna ecosystem. J Ecol. doi:10.1111/1365-2745.12059

Chapter 2
Competition for Limiting Resources, Hamilton's Rule, and Chesson's R^*

> Competition is an interaction between individuals, brought
> about by a shared requirement for a resource in limited supply,
> and leading to a reduction in the survivorship, growth, and/or
> reproduction of the competing individuals concerned.
>
> Begon et al. (1990)

> Social evolution is facilitated in proportion to the coincidence of
> fitness interests experienced, through sociality, by the
> component sub-units (partners). Such a coincidence may come
> about through two basic methods, namely shared genes
> (relatedness) or shared reproductive fate.
>
> Bourke (2011)

> The expected evolution obeys an adaptive topography defined by
> the long-run growth rate of the population. The expected fitness
> of a genotype is its Malthusian fitness in the average
> environment minus the covariance of its growth rate with that of
> the population.
>
> Engen et al. (2009)

Abstract This chapter links Chesson's R^* with inclusive fitness theory, arguing that competition for limiting resources within and between groups underlies both formulations. Chesson's R^* determines the strengths of interspecific compared to intraspecific competition, the balance of which is determined by the species having the highest rate of increase when conditions are at their worst. The latter formulation is generalized to the within- and between-group levels, l^*_{within} and $l^*_{between}$, where types compete for the lowest l^* values in the most severe regimes. Conditionally, entities with the highest growth (group, population) or reproductive (types) rates are, theoretically, the superior or dominant types. It is argued that l^* values are linked to Hamilton's rule via a formulation advanced in 2002 showing when kin should remain in groups and when they should leave, states determined by intensities of within-group compared to between-group competition.

Keywords Hamilton's rule · Inclusive fitness theory · Direct reproduction · Indirect reproduction · Ecological constraints · Hebbian "decisions" · Chesson's R^*

C. B. Jones, *The Evolution of Mammalian Sociality in an Ecological Perspective*,
SpringerBriefs in Ecology, DOI 10.1007/978-3-319-03931-2_2,
© Clara B. Jones 2014

Immediate or future consequences of fitness-maximizing "decisions" will often be unpredictable and uncontrollable, even in stable conditions, defined as persisting benefits from relatedness or from shared reproductive interests (Bourke 2011). For example, if one type dies or locates a mate of higher quality, fitness optima of Actor, Recipient, or the offspring of either type may change. Large-brained mammals should be selected to utilize a set of "decision rules" about when, under what conditions, and how to help a conspecific. Ceteris paribus, male helping will limit male mates, increasing the intensity of competition among females for mates; female helping will limit availability of female mates, increasing competition among males for mates; and, helping by both sexes will yield intense competition within both sexes for mates. An organism's "decisions" are expected to be influenced by factors in addition to self-interest, such as predation, resource dispersion, and interspecific competition (Fig. 2.1). The latter effects may, at some "decision" points, overwhelm self-interest so that "ego" loses or fails to gain inclusive reproductive benefits. Bourke (2011) stressed that benefits parameterized in Hamilton's rule are in the form of offspring produced in the future, and that the outcomes "are effects at the level of the entity (e.g., cell, organism, "type") performing the behavior, which is why they are expressed in terms of offspring number." The current section links inclusive fitness theory with competition theory as presented by Chesson's (Chesson's 2000; Chesson and Kuang 2008; also see, Amarasekare 2003, 2009; Amarasekare et al. 2004) treatments of "competitive coexistence." These two traditions are connected via the reproductive consequences to types from competition within and between groups for limiting resources convertible to direct or indirect reproduction (West-Eberhard 1975), where competition is a function of genotype × environment (including social environment) interactions in a type's thermal zone.

2.1   West et al. (2002) Places Hamilton's Rule in the Context of Intraspecific Competition

As Bourke (2011) made clear, Actors will generally prefer to facilitate the reproductive interests of kin since such choices will represent the highest likelihoods of increasing the representation of an allele in a population. However, variations in resource dispersion and quality change a type's costs and benefits and, possibly, response thresholds and hierarchies. An extension of Hamilton's (1964) treatment was advanced by West et al. (2002), showing that where within-group (local or "patch") competition exceeds (some threshold level of) between-group ("global" or population) competition, selection will favor phenotypes dispersing from natal groups, thereby, avoiding direct, genetically deleterious, competition with kin whose ecological requirements are most likely to overlap. West et al. (2002) presented a model predicting conditions under which individuals should (weak within-group competition relative to strong between-group competition) and should not (strong within-group competition relative to between-group competition) cooperate with or demonstrate altruism toward kin. This formulation takes into account all agents affected by an act, not only the relatives of an Actor (Hamilton's rule: Fig. 1.2).

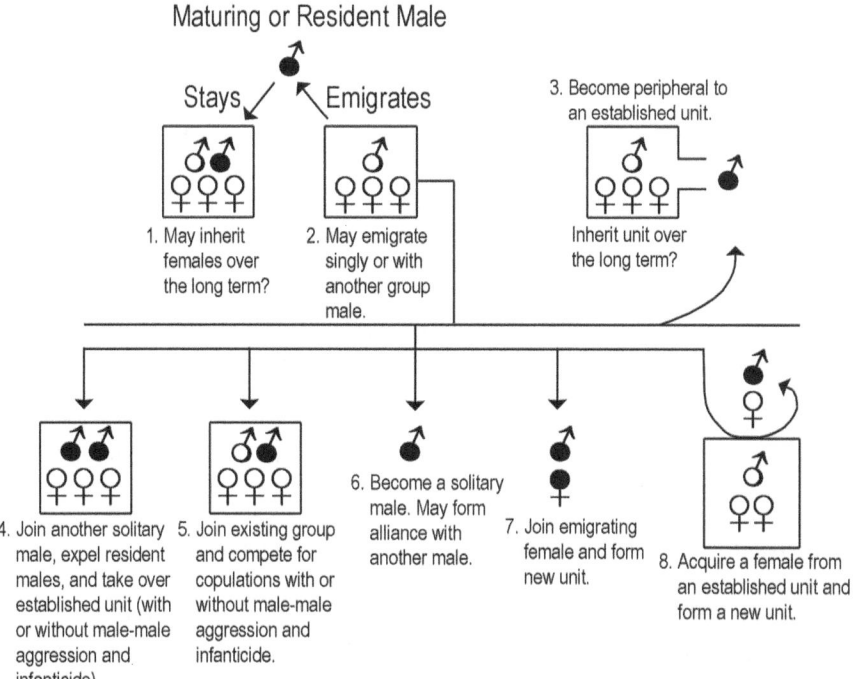

Fig. 2.1 Selected alternative reproductive "decisions" (Chap. 3) available to mammalian males of one species (*Alouatta palliata*, based on real data: Jones 2005) remaining in or after emigration from natal or resident groups. "Decision" *6* may lead to a "solitary" ("sexually segregated") population structure, and options *2, 7,* and *8* may lead to varieties of polygyny or to monogamy in which a single reproductive male (related or unrelated) monopolizes one or more than one reproductive female. Strategies *1, 4,* as well as *5* affect multimale–multifemale groups requiring male–male tolerance, possibly, leading to sociality (e.g., alliances, coalitions, dominance hierarchies or other forms of aggressive and/or reproductive restraint, possibly imposed by dominants). Other potential options are not represented in this figure (e.g., colonization). All alternative reproductive "decisions" should occur in response to condition-dependent, within- and between-group competition for limiting resources (e.g., for group membership, for mates: see Jones et al. 2008), and the potential for social evolution will be a function of differential reproductive costs and benefits in association with shared reproductive interests with kin or non-kin. Hystricognath rodents would provide a large and various group for tests of the previous topics, including the role of adult sex ratios in groups and populations as determinants of differential reproductive tactics and strategies expressed by adult females and males. In sum, abiotic and biotic characteristics of "local" and "global" regimes, including conspecific and contraspecific interactions, will determine condition-dependent "decisions" made by types. © Clara B. Jones

West et al. (2002) extended Hamilton's rule so that $r_{xy}b - c - r_{xe}d > 0$, with r_{xy} representing an Actor's relatedness to an act's Recipient (Hamilton's r), r_{xe}, an Actor's relatedness to individuals whose fitness is decreased by the act, and d represents the positive or negative effects of the act for within-group competition. In the expanded formulation, b and c are defined as in Hamilton's equation. If d increases as a result of an act, fitness may be negatively influenced, whereas, if d decreases as a result of an act, fitness may be positively influenced (relative to values of r_{xy} and r_{xe}). Thus, when applied to group maintenance subsequent to the formation of groups, a "decision" to cooperate or otherwise help a member of an aggregation or existing group will be a function of a helper's potential benefits and costs from the consequences of helping for the inclusive fitness of relatives and non-relatives in the aggregation. r_{xe} will be associated with local (single patch) competition, increasing as within-patch (density-dependent) competition for limiting resources intensifies. The extended formulation shows that helping is not inevitable, but that it is a function of its effects upon all affected types in an aggregation (see Sect. 2.2). This model, thus, demonstrates how groups of related or unrelated types might arise from aggregations, possibly forming stable maintenance of groups given constraints imposed by Hamilton's rule. Furthermore, like Hamilton's original treatment (1964), the parameters advanced by West et al. (West et al. 2002; Rodrigues and Gardner 2012, 2013) are measurable or can be estimated in Nature; thus, technically, they are testable with further theoretical treatments and by field and laboratory experimentation. The extensive dataset on within- and between-group interactions of polygynandrous African lions (*Panthera leo*) assembled by Mosser and Packer (2009) may provide tests of the previous model.

Crespi's (2006) discussion of West et al. (2006) applies, as well, to West et al. (2002) by emphasizing the necessity to consider the spatial scale (environmental context) of competition and its effect upon the differential costs and benefits of cooperation for individuals of a population. As Crespi (2006) pointed out, the dispersion of limiting resources may "create opportunities to compete and to cheat." The social potential of a given "patch," in theory, should be measurable. The previous author considered the spatiotemporal distribution and abundance of limiting resources hypothesized to favor groups. A fundamental theoretical problem has been explaining how cooperation is sustained once it is expressed since the benefits of cheating and self-interested behavior are expected to outweigh the benefits of cooperation or altruism (Maynard Smith 1974) since Actor is related to itself by an r of 1.00. West et al. (2006) showed that the "spatial scale of competition can drive evolutionary dynamics of social interactions among non-kin," conditions hypothesized to be a function of whether competition is local or global. Emphasizing the relative nature of inclusive-fitness maximizing, Crespi (2006) expressed the (spatial) ideas of West et al. (2006) schematically, as follows:

- Cooperation Local, Competition Global $\rightarrow$ Local Cooperation Favored by Selection
- Competition Local, Cooperation Global $\rightarrow$ Local Competition Favored by Selection

where Local implies competition for limiting resources within "patches" or within groups, and Global implies competition in the population at large (between groups). The treatments of West et al. (West et al. 2002, 2006; Rodrigues and Gardner 2012,

2013), thus, expand Hamilton's (1964) work by defining the population-level contexts of cooperation and altruism and conditions under which those responses by Actors should conform to Hamilton's rule and be favored by selection.

2.2 Generalizing Chesson's $R*$ and Linking it to Hamilton's Rule

The logic of "competitive coexistence" theory holds that where two species are in competition for the same resource, the dominant species will have the higher rate of increase "at the lowest resource level" (Stearns 1989; Schoener 1974). Discussing the evolution of stable coexistence among species, Chesson (Chesson 2000; Amarasekare 2003, 2009; Amarasekare et al. 2004) proposed that mechanisms of coexistence ("stabilizing mechanisms") involve (condition-dependent) facilitation of an inferior species by a superior species, classifying promoters of coexistence as mechanisms of resource partitioning, predation ("natural enemies"), and spatiotemporal fluctuations in population densities (e.g., along environmental gradients). Chesson (2000) pointed out that (condition-dependent) facilitation via "stabilizing mechanisms" promotes "long-term per capita growth rate" of inferior species by facilitating an increase in the density (rate of increase, growth rate) of the inferior species, decreasing stochasticity in their population numbers.

Relying upon previous quantitative treatments, Chesson (2000) argued that mechanisms of coexistence reflect "direct competition" and "resource dynamics" where species are limited by a single resource. Such conditions are governed by the $R*$ rule whereby $R*$ "is the resource level at which a species is just able to persist." The previous author goes on to note that, "The winner [sic] in competition is the species with the lowest $R*$ value," comprising the relative potential for types in low-density conditions to survive and reproduce. The same author pointed out that the ability to persist at low resource levels can be partitioned into causes such as success of foraging strategies and rates of predation. The potentially destabilizing effects of interspecific competition can be kept in check via mechanisms of coexistence (e.g., facilitation), yielding a condition whereby interspecific competition would be, ceteris paribus, less intense than intraspecific competition.

It follows logically that intraspecific competition applies to direct competition between groups and between types in groups where (condition-dependent) niche space of a type overlaps with niche space of other groups. Social biology, then, pertains to mechanisms whereby types manage direct competition with other types via interindividual actions (Table 1.1, Fig. 2.2). By analogy with Chesson's (2000) formulations, a dominant type is the one with the highest growth rate when a resource is sparsely distributed or least abundant (= low resource dispersion), and facilitation by a dominant or subordinate type are expected to increase the low growth rate of a subordinate type, relative to local ("patch") or "global" (population) regimes (Table 2.1). It follows that, because types within groups are more likely to be similar

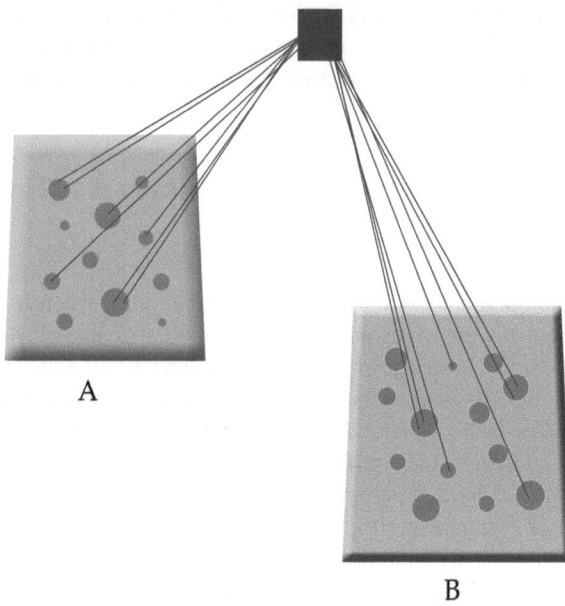

Fig. 2.2 A "nested vision," 3D map, conceptualizing a type (*black square*) interacting with two environments, *A* and *B* (e.g., local regimes, "patches," habitats, or different regions of a gradient). In each landscape, each *line* represents a trade-off of two traits, *x* vs. *y*, whereby a trade-off indicates that expression of one or a suite of traits decreases energy available for allocation to traits associated with survival, reproduction, or growth. Sizes of different circles (*dark grey*) represent differences in relative reproductive success of each trade-off, and performance in each patch has different optimal tradeoffs determined by condition-dependent features. Alternatively, each *dark grey circle* may be envisioned as the relative reproductive benefit gained from each trade-off (each line, *x* vs. *y*) when interacting with two group members, *A* and *B* ("social competition": Crook 1972; West-Eberhard 1979). Imagine that the trade-offs are survival (*x*) and reproduction (*y*) of two social traits where *x* = care of one's own offspring, *y* = "helping" the reproduction of another group member. In "fine-grained" conditions (environmental changes less than generation time), single optima will be favored in both locales (*A* and *B*), a scenario descriptive of the reproductive optimum of many large mammals. In other conditions, specialization to the separate locales will obtain ("coarse-grained" conditions: environment changes occur longer than generation time, characteristic of many small mammals). In other regimes, polymorphism is selected, for example, in predictable, heterogeneous environments such as seasonal regimes.

Changes in values of trade-offs across each performance curve (single optimum trade-off, specialization to *A* and *B*, or polymorphism) represent responses to competition within groups (Maharjan et al. 2013; Adler et al. 2013). These changes may be detected by differential interaction rates of types. Thus, these relationships should be testable. Intensity of within-group competition relative to between-group competition will determine relative costs and benefits to types (in this case, a type expressing traits *x* and *y*) of residing with kin or with unrelated types. Where groups of reproductive kin coreside, altruism may be favored (Fig. 1.2). Where groups of unrelated reproductive types are favored, social evolution will depend upon shared reproductive interests ("fates") of group members, constraining evolutionary potential for the rise of altruism but with the capacity for the rise of cooperation in the Hamiltonian sense (Bourke 2011). For reasons that are poorly understood, social evolution has been retarded in most mammalian taxa, even where groups of related reproductive types coreside, possibly because competitive regimes have disfavored coresidence of kin (i.e., where within-group competition is intense compared to between-group competition: Chap. 2). © Clara B. Jones

Table 2.1 Assuming two types (a "dyad") compete for a limiting resource, three possibilities obtain to manage competition: Positive, Neutral, or Negative interactions. Each of the three mechanisms has the potential to induce group-formation or, in some conditions, to maintain groups once they have formed (see examples). The evolution of sociality (Fig. 1.2) may arise initially *via* facilitation (e.g., daughters helping mothers), *via* tolerance (e.g., mothers tolerating presence of daughters), or *via* inhibition (usually) *via* a superior type repressing reproduction of an inferior type (e.g., mothers repressing the reproduction of daughters). Care must be taken to precisely identify whether only one type in a dyadic interaction facilitates the reproduction of the other type or whether both types facilitate reproduction of the other type in the dyad. © Clara B. Jones Based on Begon et al. (1990)

Potential responses to limiting resource available for monopolization

Mechanism employed by focal type (Actor) to manage competition with recipient and/or recipient's kin	Effect of interaction on recipient and/or recipient's kin	Description of response
Facilitation (positive interactions)	+	Contest competition or exploitative competition initially; only certain types in group capable of monopolizing resource
		Management of competition with Recipient increases Hamiltonian fitness of Actor and Recipient or increases Recipient's Hamiltonian fitness at expense of Actor's, e.g., some avoidance responses; restraint by Actor; change in status or rank; dispersal; death; cooperation, altruism; reciprocity, alliances, coalitions; increase or decrease thermal tolerance (niche space), possibly, via roles or division of labor (intermediate or high levels of competition or stress)
Tolerance (neutral interactions)	0	No attempt to manage competition with Recipient or no competition, e.g., aggregation?; some avoidance responses; standoff, "cold war," détente; "social facilitation" may be precursor to facilitation or inhibition
Inhibition (negative interactions)	−	Management of competition increases or does not change Actor's Hamiltonian fitness but decreases Recipient's Hamiltonian fitness, e.g., "parental control"; "parental manipulation"; ejection; repression of reproduction or other selfishness by Recipient or Recipient's kin (low or high levels of competition or stress?); the role of negative interactions requires close inspection for its ability to induce "altruism" ("helping") in potential helpers ("donors"; see this brief, Coda)
		Older types in good condition may persist in some groups until types in poorer condition disperse or are ejected or die
		Younger or older types in good condition may inhibit types in poor condition

Table 2.1 (continued)

Potential responses to limiting resource available for monopolization

Mechanism employed by focal type (Actor) to manage competition with recipient and/or recipient's kin	Effect of interaction on recipient and/or recipient's kin	Description of response
Facilitation, tolerance, or inhibition	$+$, 0, or $-$	Management of competition may change group competition and/or mean levels of competition in group
Facilitation and tolerance	$+$ and/or 0	Focal type may facilitate Recipient by mere presence
Tolerance and inhibition	0 and/or $-$	Any type in group capable of monopolizing resource ("scramble" competition initially"?)

than types between groups, their niche spaces will be more likely to overlap. Thus, ceteris paribus, within-group competition will be more intense than between-group competition, and the differential reproductive costs and benefits to types within groups will determine, in part, whether a type "decides" to stay or to leave its group of origin. This scenario, based upon West et al. (2002), may explain, in part, why most mammals are "solitary"; however, the latter authors do not parameterize within-group factors that may lead to coresidence of kin or unrelated types. The schema previously formulated assumes that inter- and intraspecific assemblies arise from cumulative effects of the reproductive rates of types (conspecifics) whose reproductive interests overlap. This "transference" from a higher level of organization to another, lower, scale demands quantitative, including experimental, testing. The emphasis herein highlights the similar ways in which phenomena at each scale function.

Herein, I use notation, l^* (level), derived from Chesson's R^*: $l^*_{between}$ and l^*_{within} where $l^*_{between}$ is defined as the lowest resource level at which a type's mean reproductive rate continues to increase when its resident group competes with other groups composed of an overlapping or nonoverlapping set of types. l^*_{within} is defined as the lowest resource level at which a type's mean reproductive rate continues to increase in competition with other types in its resident group. Thus, the type with the lowest $l^*_{between}$ or l^*_{within} is the dominant (superior) type, *in a condition*. Since the expression of social actions (whereby one type facilitates the reproduction of a conspecific type) is density dependent, since social actions directly impact rates of increase or decrease of individuals, groups, and populations, and since Hamilton's rule is fundamental to the operation of social actions, inclusive fitness theory is inherently tied to the competition of individuals for limiting resources, to competition for the lowest $l^*_{between}$ and l^*_{within}, and to the evolution of sociality, herein, the evolution of mammalian sociality.

Groups are composed of types and a type's discrete actions will affect within- and between-group competitive regimes. The actions of a social type may decrease intensity of within-group competition, on average, as long as the response is not neutralized or outweighed by another type's counterstrategy (e.g., rejection of copulation). In mammals and some other vertebrates, elaborate signals and displays are components of, often lengthy, behavioral sequences serving to increase likelihoods that a social action will more than return its energetic (reproductive) investment to the Actor. The relevant point is that social traits expressed by types will determine whether intensity of competition is higher within or between groups. In the heterogeneous regimes in which mammalian evolution occurred, there would likely have been a dynamic and shifting balance of within- and between-group intensities with consequences for $l^*_{between}$ and l^*_{within}, for patterns of immigration and emigration of types, as well as for the potential for cooperation or altruism between types.

After Chesson's (2000) treatment, social mechanisms may stabilize interactions between and within groups by decreasing stochasticity of one or more type's reproductive rates since $l^*_{between}$ and l^*_{within} impact the reproductive potential of other types. It must be kept in mind, however, that a social act that stabilizes one or more types within or between groups will not necessarily be beneficial to all types concerned (e.g., "desperados," social parasites, the spiteful). Obviously, then, it would seem unlikely to find a group of pure altruists, since there seems always some benefit of retaining genes designed to repress the selfishness of others. As already stated, "decisions" made by types and the consequences of these decisions for resources and for other types (competitors) are limited by the parameters of Hamilton's rule (Bourke 2011; Lehmann and Keller 2006; Frank 2013). West et al. (2002) showed that Hamilton's rule reflects "decisions" available to focal types (Actors: Table 1.1) in ecological regimes (Figs. 1.1 and 2.1, Table 2.1), determining differential, condition-dependent reproductive costs and benefits to types (Fig. 1.2). "Decisions" obtain where types compete for limiting resources in their thermal zones.

References

Adler PB, Fajardo A, Kleinhesselink AR, Kraft NJB (2013) Trait-based tests of coexistence mechanisms. Ecol Lett. doi:10.1111/ele.12157
Amarasekare P (2003) Competitive coexistence in spatially structured environments: a synthesis. Ecol Lett 6:1109–1122
Amarasekare P (2009) Competition and coesixtence in animal communities. In: Levin SA Princeton guide to ecology. Princeton University Press, Princeton
Amarekare P, Hoopes MF, Mouquet N, Holyoak M (2004) Mechanisms of coexistence in competitive metacommunities. Am Nat 164:316–326
Begon M, Harper JL, Townsend CR (1990) Ecology: individuals, populations, and communities, 2nd edn. Blackwell Scientific Publications, London
Bourke AFG (2011) Principles of social evolution. Oxford University Press, Oxford
Chesson P (2000) Mechanisms of maintenance of species diversity. Ann Rev Ecol Syst 31:343–366
Chesson P, Kuang JJ (2008) The interaction between predation and competition. Nature 456:235–238
Crespi BJ (2006) Cooperation: close friends and common enemies. Curr Biol 16:414–415
Crook JH (1972) Sexual selection, dimorphism, and social organization in the primates. In: Campbell B (ed) Sexual selection and the descent of man. Aldine, Chicago, pp. 1871–1971

Engen S, Lande R, Sæther B-E (2009) Reproductive value and fluctuating selection in an age-structured population. Genetics 183:629–637

Frank SA (2013) Natural selection. VII. History and interpretation of kin selection theory. J Evol Biol 26:1151–1184

Hamilton WD (1964). The genetical theory of social behavior. J Theor Biol 7:1–52

Jones CB (2005) Behavioral flexibility in primates: causes and consequences. Springer, New York

Jones CB, Milanov V, Hager R (2008) Predictors of male residence patterns in groups of black howler monkeys. J Zool 275:72–78

Lehmann L, Keller L (2006) The evolution of cooperation and altruism—a general framework and a classification of models. J Evol Biol 19:1365–1376

Maharjan R, Nilsson S, Sung J, Haynes K, Beardmore RE, Hurst LD, Ferenci T, Gudelj I (2013) The form of a trade-off determines the response to competition. Ecol Lett 16:1267–1276

Maynard Smith J (1974) The theory of games and the evolution of animal conflicts. J Theor Biol 47:209–221

Mosser A, Packer C (2009) Group territoriality and the benefits of sociality in the African lion. Anim Behav 78:359–370

Rodrigues AMM, Gardner A (2012) Evolution of helping and harming in heterogeneous populations. Evolution 66:2065–2079

Rodrigues AMM, Gardner A (2013) Evolution of helping and harming in viscous populations when group size varies. Am Nat 181:609–622

Schoener TW (1974) Resource partitioning in ecological communities. Science 185:27–39

Stearns SC (1989) Trade-offs in life-history evolution. Funct Ecol 3:259–269

West SA, Pen I, Griffin AS (2002) Cooperation and competition between relatives. Science 296:72–75

West SA, Gardner A, Shuker DM, Reynolds T, Burton-Chellew M, Sykes EM, Guinee MA, Griffin AS (2006) Cooperation and the scale of competition in humans. Curr Biol 16:1103–1106

West-Eberhard MJ (1975) The evolution of social behavior by kin selection. Q Rev Biol 50: 1–33

West-Eberhard MJ (1979) Sexual selection, social competition, and evolution. Proc Am Phil Soc 123:222–234

Chapter 3
Flexible and Derived Varieties of Mammalian Social Organization: Promiscuity in Aggregations May Have Served as a Recent "Toolkit" Giving Rise to "Sexual Segregation," Polygynous Social Structures, Monogamy, Polyandry, and Leks

It seems very improbable that any mammal is ever truly socially neutral; individuals of other species may be disregarded, but to have no response, either positive or negative, to conspecifics, seems inherently improbable.

Ewer (1968)

All evolutionary processes must work by modification of existing systems, so in this sense, it is expected that social traits would be foreshadowed in traits of solitary ancestors.

Bourke (2011)

Since the marsupials have radiated to fill a wide variety of ecological niches, and since this radiation has been accomplished utilizing the same fundamental body plan, the marsupials are the only "control" group with which we can test hypotheses about the evolution of behavior within the eutherian mammals.

Eisenberg (1981)

Abstract Chapter 3 argues that extant mammals are characterized by an ancient social "toolkit" derived from the traits of ancient group-living mammals. An important lesson highlighted by a review of extant social evolution is that animals in heterogeneous regimes are not necessarily group living, although, extreme environments (sublethal stress?) appear to favor higher grades of sociality. A review of the literature suggests that flexible social structures evolved from "promiscuous" aggregations of reproductive males and females characterized by nonoverlapping ranges and that body sizes, home-range sizes, and male–male tolerance are driven by evolution in thermal ("patch") regimes.

Keywords Social architecture · Abiotic factors · Biotic factors · Thermal niche · Sexual dimorphism · Promiscuity

I assume in Chaps. 3, 4, and 5 that types "decide" to join or to leave groups or that they are expelled from groups as a function of their l^* values and that these values reflect different viabilities under different thermal regimes (Chap. 2). Reaction norms can be partitioned into causes comprising traits covarying with fitness (egg to adult viability: Ketola et al. 2013), and, for students of mammals, we are fundamentally interested in the performance of traits in fluctuating regimes (Fig. 2.2; Jones 2009). In addition to reviewing factors inducing group structures in mammals, the present brief concerns the varieties of mechanisms employed by group-living mammals to manage competition (Table. 1.1, 2.1) and to gain a reproductive advantage over conspecifics, seeking to minimize $l*_{within}$ and/or $l*_{between}$. Throughout this and the following section's discussion of mammalian population architectures, it is useful to ask: How might social actions function to minimize both of the latter functions, and what condition-dependent reproductive costs, benefits, and tradeoffs attend each fitness-maximizing act (Fig. 2.2)? The statement by Eisenberg (1981) quoted at the beginning of this chapter informs the reader that marsupials may be employed as a control group for questions related to mammalian evolution. In effect, the previous author's view is that marsupial morphology is a relatively invariant constant against which traits characterizing other mammalian taxa may be compared. Though marsupials are an evolutionarily primitive mammalian group (Metatheria), they occupy a broad range of environmental regimes, exhibiting many types of population structure found in the class as well as several examples of "fast" life-history trajectories (Stearns and Koella 1986), similar to most small mammals (Eisenberg 1981). As a result of Eisenberg's (1981) influence, marsupials, in particular, macropods, are treated relative to eutherians herein. A "toolkit" is proposed (Table 3.1, Fig. 3.1) whereby proteins associated with tolerant or facilitating phenotypes are available to connect and reconnect like ©Lego pieces, differently colored pieces comparable to different proteins in the toolkit. Mammals, also, are characterized by relatively large brains controlling and coordinating action patterns embodying noteworthy abilities for opportunistic, facultative "decision making."

Information is presented taxonomically, not, chronologically. Synthesis of Table 3.1 suggests that aggregations (e.g., during opportunistic foraging or hunting) and social tendencies or sociality among "prehistoric" mammals are associated (1) with herbivory and/or carnivory; (2) with spatiotemporal dispersion of limiting resources, particularly, food, water, and breeding sites; (3) with predation pressures and other associates of competition; (4) with indicators of sexual selection; (5) with variations in geochemical events (e.g., climate); and, (6) with morphological design ("phylogeny"). These and other early mammalian features may have constituted a "toolkit" of genetically correlated traits that, when combined and recombined, were favorable to social evolution. Patterns of events detected in the table also suggest that costs associated with detection, search, acquisition, or allocation of limiting resources were correlated with dangerous, difficult, rare, or risky conditions. Information summarized in this table indicates that large body size and, probably, a generalized phenotype, disfavored the expression of quasi-social or social traits, partially explaining why sociality is restricted in most orders of class Mammalia.

Table 3.1 Table summarizing social traits in "prehistoric" social mammals. (Table and inferences based on Kielan-Jaworowska et al. 2004; Turner 2004; Kermack and Kermack 1984; Eisenberg 1981; Chapman and Feldhamer 1982; Brook and Bowman 2002; Morales and Giannini 2013; Rakotoarisoa et al. 2013; Wilson et al. 2012; Packer 1986; KD Angielczyk, personal communication)

Identity	Food category	Social trait(s)	Notes
Marsupials: *Thylacines*	Carnivore	Cooperative hunters? (opportunistic?)	Cooperative hunting may have been favored to subdue "giant" kangaroos and "hippo-sized" wombats; *Pucadelphys andinus*, a Bolivian type from the early Palaeocene, is the earliest social taxon so far described (Ladevèze et al. 2011; see text)
Kangaroos	Browsers (herbivores)	Foraging in groups (opportunistic?)	May be possible to estimate what kangaroos and other browsers ate and the spatiotemporal dispersion of these resources that might predict likelihoods of grouping; grouping (opportunistic?) may have been favored in response to predation; large body size would have reduced effects of environmental heterogeneity
Proboscids (n.b. classic cases of convergent evolution in this group): Deinotheres (elephants)	Browsers (herbivores)	Group foraging, possibly in response to reduced costs of searching for and partitioning food sources	Similar to extant elephants, young very vulnerable to predation, apparently, favoring cooperation among females; adults lack tusks, possibly increasing costs from predation despite large body size; sexual segregation (all-male herds) multi-level societies may have been favored by limited supplies of water, as seems to have occurred among some anthropoids, including, humans (Sect. 6)
Gomphotheres (mastodons)	Browsers (herbivores)	Same as above	This group gave rise to extant elephants; tusks present (defense?, sexual selection?); other factors same or similar to above
Mammutidae ("true mastodons," not mammoths)	Browsers (herbivores); dentition consistent with adaptation(s) to abrasive food, suggesting a broadening niche and increased environmental stress, possibly in response to competition from other herbivores, particularly, elephants, and/or from omnivores	Same as above	Extravagantly elongated tusks, consistent with sexual signaling; similar to saber-tooth tigers, the elaborate display probably evolved in response to sexual selection and may have compromised survival; vulnerable to hunting by human hunting; other factors same or similar to above

Table 3.1 (continued)

Identity	Food category	Social trait(s)	Notes
Elephantids ("true elephants")	Browsers (herbivores); dentition consistent with adaptation to "rough" vegetation, suggesting increasing environmental stress and competitive pressures	Same as above	Tusks; largest of any elephant, suggesting that constraints from energy requirements (endothermy and large body size) may have hastened its extinction (0.5 mya; see other notes for proboscids)
Mammuthus (mammoths, "elephantids")	Browsers (herbivores); see comments above re: dentition and abrasive food (but, note characteristic "parallel ridges" for "wooly mammoths")	Same as above	Large, heavy, tusks (see Mammutidae, above); thermoregulatory pelage; huge body size and consequently large food requirements (see other notes for proboscids); some information about this group gleaned from cave paintings
Primates; Anthropoids (monkeys and apes): Adapidae (51 mya), Cercopithecidae (3.5 mya); Apes (17 mya); *Australopithecus* (3.5 mya); *Paranthropus* (2.5 mya); *Homo erectus* (1.8 mya); *H. antecessor* (780,000 ya); *H. neanderthalensis* (200,000 ya)	Herbivores, folivores, frugivores, or omnivores	Group-living favored by energetics of food search and acquisition; opposable thumb facilitates food selectivity (associated in browsers with sociality and higher levels of plant toxicity (food stress) compared with grazers); n.b. forest living not amenable to grazing; opposable thumb; also, favors evolution of grooming, the most common social action pattern among nonhuman primates; *Paranthropus*: morphological and dental characteristics relatively specialized for "rough," hard to process, and gritty food (see "Proboscids" above); unclear whether *Paranthropus* used tools; fire- and tool-use confirmed for *H. erectus* (Sect. 6); *H. erectus* may have hunted cooperatively; *H. antecessor*: fire- and tool-use; was, probably, cannibalistic; tools, manufacture, industry highly developed in *H. neanderthalensis*, but presence of language debated; relative brain sizes increased as social traits proliferated	Many anthropoids medium sized, vulnerable to predation, limited by water (see Deinotheres above); some extinct forms "rodent-like," "lemur-like," prominent canines in some families (see above, "tusks"); *Paranthropus*, marked sexual dimorphism in body size; *H. erectus*, exhibiting robust traits, earliest known, direct ancestor of *H. sapiens*; *H. erectus* a scavenger, possibly, a hunter; *H. antecessor* oldest human species found in Europe (see Sect. 6, *H. sapiens*, 160,000 ya)

Table 3.1 (continued)

Identity	Food category	Social traits(s)	Notes
Bats	Frugivores (Megachiroptera); primarily, insectivores (Microchiroptera)	Extant frugivorous bats "roost gregariously;" however, the issue of whether extant Megachiroptera "forage away from their roosts" is an "open question" (TH Fleming, personal communication)	Megachiroptera habitat distribution comparatively "restricted," probably related to competition from birds and patterns of fruit tree dispersion; extant frugivorous primates and bats are polygynous (THF, p.c.); bats not well known (THF, p.c.)
Creodonts	Carnivores	Probably (opportunistic?) scavengers	Feet not adapted for running to capture prey; did not compete with haenids of this time because the latter were insectivores; evolution to hunting limited where legs short relative to body-, shoulder-, and head-size; probably responsible for extinction of Mesonychians (see below)
Carnivores: seals and relatives	Carnivore	?	Many extant pinnipeds "highly gregarious"
Cats (*Smilodon*, "saber-tooth" cats)	Carnivore	Early cats were probably "ambush" predators, but *Smilodon* probably lived and (opportunistically?) hunted in groups, similar to lions	Canine teeth extremely exaggerated (see notes above); (Did strong sexual selection favor sociality?); limb bones relatively short (how is allometry related to the evolution of cooperative hunting?); present-day lions are strongly sexually selected and young suffer severe levels of infant mortality (see elephants, above)
Hyenas	Carnivore (early forms insectivores)	Modifications of the jaw (~ 5 mya) permitted scavenging, possibly, some type of coordinated hunting	How were scavenging, kleptoparasitism, and opportunistic hunting related to evolution of hunting and of sociosexual systems (n.b. extant hyenas)?
Cloven-hoofed mammals (ungulates): Entelodonts (related to Artiodactyls)	Carnivores	Scavengers and opportunistic or primitive hunters?	Allometry would have constrained hunting; exaggerated facial structures suggest sexual selection

Table 3.1 (continued)

Identity	Food category	Social traits(s)	Notes
Mesonychians (Artiodactyls)	"Wolves on hooves," carnivores (first major mammalian predators)	Coordinated or group hunters	Extinction due to climate change and competition from creodonts; all ungulates adapted to eat "lower-grade food" and have "reduced teeth"
Pigs	Earliest suids strongly herbivorous; later suids, omnivores	Coordinated or cooperative foraging?	Males bear exaggerated tusks and facial structures; young highly vulnerable to predation (see elephants and *Smilodon*, above); pigs have high reproductive rates compared to other ungulates; terrestrial ungulates mostly herbivores, some grazers
Perissodactyls (later horses: *Equus stenonis*)	Grazer	Generally depicted as group living because dental diagnosis zebra-like	Grazers generally not highly selective foragers compared to browsers, and grazers less likely to be group living; however, *Equus* teeth like those of zebras rather than "true horses" were "supremely efficient grazers; thus, possibly, selective grazers and social equids

It is important to keep in mind that because of variations in "tempos and modes" of biogeochemical events across historical time periods, and, because of variations in the contents of social "toolkits," patterns of sociality may not be straightforward or lineage-specific. Furthermore, convergent evolution is almost certainly an important factor to consider for an understanding of mammalian social evolution (Chaps. 3–5), as it is for social insects (Woodard et al. 2011). Sociality most likely evolved more than once in the Class, with similarities within and between genera accounted for by eco-morphological constraints. Other evolutionary trends are detected in the table that require systematic investigation, such as the observation that many prehistoric felids, typically, "solitary" carnivores, may have exhibited opportunistic, or, "true," scavenging or hunting, or, kleptoparisitism, a relatively common form of "social parasitism" among mammals (Jones 2005). Numerous taxa in this table are unresolved or poorly described; as a result, dates and timelines are subject to modification. Systematic research on "solitary" and "social" prehistoric omnivores is needed since omnivorous habits may be destabilizing (Gellner and McCann 2012), possibly influencing patterns of diversification or extinction.

Fig. 3.1 A prehistoric dog, *Hesperocyon gregarius* (Canidae, Hesperocyoninae), endemic to North America, 37–31 mya. These "fox-like," carnivores were probably communal, "stalking and pouncing" small animals. Their relatives, the hypercarnivorous, possibly omnivorous, Borophaginae, gave rise to extant wolves, foxes, coyotes, jackals, and dogs. (©Victoria Wheeler)

3.1   The Evolution of Thermal Niches and the Evolution of Mammalian Sociality

Intra- and inter-type effects via differential thermal niches within and between local regimes determine population structure. "The factors that influence space use in female mammals ultimately determine social organization" (Fisher and Owens 2000; also see Jarmon and Southwell 1986) "because females ultimately limit male reproductive success" (Emlen and Oring 1977; Trivers 1972). Mammalian "social" organization varies by subclass, with eutherians more sensitive to dispersion of resources, while macropods are more sensitive to climate (Fisher and Owens 2000). Perhaps for the latter reason, the "body plan" and population architectures of macropods and other marsupials has remained conserved and stable over time, reflecting adaptation to "moving target" environments that cannot be tracked by types (Roughgarden 1979) within the constraints of generation time (Jones 2012). For similar reasons, rodents are a model system for eutherians as a whole.

Among mammals, larger home range sizes correlate with larger group sizes, and larger groups are usually "social" (Fisher and Owens 2000), although cooperatively breeding mammals and the long lived, eusocial mole rats (Bathyergidae) are small taxa in small groups. The latter observations imply that more than one "route" to sociality may have influenced mammalian social evolution (Chap. 1). The evolution of large groups may be opposed by "kin selection" since, according to Hamilton (1964), "decisions" by type in a group to reproduce indirectly rather than directly will favor small group size. The latter effect may lead to conflicting reproductive optima between the sexes since mammalian females, even in "solitary" species, are more likely to be social (e.g., exhibiting helping, hygienic or stress-decreasing grooming, or allomothering: see Chap. 8). This apparent trade-off between the benefits of kin selection and the benefits of living in large groups highlights a physiological dilemma (foraging efficiency) that may explain: (1) the relative infrequency of sociality in mammals as well as (2) the apparent correlation between division-of-labor and stability of limiting resources, particularly, food and refugia (Crespi 2007; Alexander et al. 1991).

Higher grades of sociality, particularly, division of labor, appear to demand a significant dedication to specialist strategies, particularly, foraging strategies and feeding selectivity, although selectivity of plant species choice and of plant parts is characteristic of large mammalian herbivores, as well. In macropods and eutherians, small taxa are more "selective foragers" than large taxa, and many group-living mammalian herbivores demonstrate a significant degree of specialization ("discriminative feeding": see Sedio and Ostling 2013; Owen-Smith and Chafota 2012) in their feeding habits (reviewed in Fisher and Owens 2000; also see Bodmer 1990; Milligan and Koricheva 2013; Di Stefano et al. 2011; Owen-Smith and Chafota 2012; Matsuda et al. 2013; Kermack and Kermack 1984, p. 12). "Selective feeding" may place limits on the evolution of traits associated with $l*_{within}$ and $l*_{between}$. Seemingly paradoxical, the transition to sociality in mammals has, also, been constrained by flexible physiological and behavioral characters in mammals, decreasing benefits of and opportunities for division-of-labor since in these, often large, species, totipotency reigns (single types perform many different tasks).

Finally, most large mammals are iteroparous breeders and most eusocial taxa, including, social insects and eusocial Bathyergids, exhibit very high reproductive rates relative to body size (see macropods for interesting cases as per Eisenberg 1981), further suggesting that cooperation in large mammals and sociality in small mammals are products of different "routes" that may be differentially energy-efficient relative to (thermal) conditions. Possibly supporting the latter view is Lacey's (2000) finding that social Bathyergids have evolved in arid habitats providing abundant, evenly distributed, subterranean supplies of food. On the other hand, Stahler et al. (2013), studying reproductive female wolves (*Canis lupus*), concluded that: "Large body size and sociality [promote] individual fitness in stochastic and competitive environments," regimes characteristic of those in which most eutherian mammals evolved (see Jones 2009). A mammalian "toolkit" (Table 3.1), then, might have permitted more than one "route" (Chap. 1) to sociality, but only if shared reproductive interests were obtained (Chap. 2).

3.2 Abiotic and Biotic "Drivers" of Body Sizes and Home-Range Sizes in Mammals

Following Fisher and Owens (2000), in both macropods and eutherians, "variation in body size was related to variation in home-range size." Habitat productivity measured by rainfall, however, was the primary effect for home-range size across macropods (negative correlation), ecological factors (e.g., food dispersion, "patchy" distribution of limiting resources), for most eutherian groups. Interspecific differences in macropod home-range size, however, were "attributed to diet," and it is important to know whether this effect is a general $R*$ function (Chap. 2) where different species compete for limiting resources. Among both macropods and eutherians, large animals (e.g., eutherian grazers: Bodmer 1990) are "much less selective foragers than small species" (but see Kermack and Kermack 1984), and "mean group size [of both taxa] is [positively] correlated with body size." The aforementioned associations highlight the primary drivers of "sociality" in the class; however, Fisher and Owens (2000) review additional trends. Based on the aforementioned associations, in general, abiotic factors are more robust predictors of "sociality" for macropods, possibly, consistent with the findings of Jetz and Rubenstein (2011) for birds. Importantly, however, the latter authors point out that climate variability may be used as a proxy for heterogeneity of food resources, a testable hypothesis for mammals and other vertebrates.

Continuing to highlight the macropod paper, Fisher and Owens (2000), also, reported that "variation in body size was related to variation in home range size," a finding consistent with findings for eutherians. Data on body size relative to abiotic and biotic factors are of particular note because body size reflects first principles of ecology (e.g., energy acquisition, consumption, and allocation). Based upon phylogenetically independent contrasts, variation in macropod home ranges was more strongly associated with habitat productivity than for eutherians, with rainforest species exhibiting small home ranges, "arid zone" taxa, the largest home ranges. The latter findings for macropods conform to findings for eutherians. Importantly, home-range size, relative to group size, reflects energy requirements and, possibly, energy reserves, thus providing opportunities to test numerous hypotheses theoretically and empirically (e.g., Smith et al. 2010), and in ecological context. There do appear to be outliers, however. For example, consistent with Bodmer's (1990) analysis, fungi-eating *Bettongia* spp. (bettongs) do not appear to follow classical patterns as per home-range metrics (see also fungus-eating Primates: Callitrichidae, Fig. 5.1).

The discussion of foraging by Fisher and Owens (2000) exemplifies the "toolkit" paradigm, advanced in the present document, as well as the characteristic flexibility of mammalian behavior and population structures. In addition, despite *caveats* pertaining to spatiotemporal models of ecology (Chap. 6), the previous authors demonstrated the explanatory power of resource, particularly, food, dispersion (distribution, abundance, and type) for edifying variations in population structure. For example, across mammals, "foraging habitat of small herbivores is patchy because they feed on more clumped and sparsely distributed food" (Fisher and Owens 2000; also, see Lee and Cockburn 1985). On the other hand, "food for larger herbivorous mammals is more patchy, because habitat is more heterogeneous at larger scales" (Fisher and Owens 2000; also, see Bodmer 1990). Thus, when assessing patterning of mammals in space and time, body size relative to resources must be considered, and these variables might vary by local conditions and by habitat, including, abiotic (e.g., soil) and biotic (e.g., tree line) gradients. Fisher and Owens (2000) highlight several extreme features of Australia's "extreme" environments that might account for the differences between macropods and eutherians, including wide variation in climate, low productivity of forests, "small range of body sizes relative to the strong climate gradient" (n.b., scale), and stasis in one or more mammalian characters (see Eisenberg 1981).

Continuing with macropod: eutherian comparisons, Fisher and Owens (2000) noted that, in both taxa, mean group size and body size are correlated. Group size and group living are associated positively in macropods and eutherians, and a predictor of variations in population organization might be the relationship between "patch" or habitat or population density and body size and/or group size. On the other hand, the highest grades of sociality are associated with genetic homogeneity caused by philopatry and/or recruitment of kin, conditions favored by "kin selection" according to the latter authors. Consistent with the aforementioned findings, variations in density were negatively correlated with female home-range size. Thus, higher densities are associated with body sizes and home range sizes in a manner that should reflect female dispersions and subsequent "mapping" of male dispersions onto those of females (Chap. 6).

3.3 What Roles Do Mammalian Males Play in Determining Population Structure? Interactions Between Intrasexual Selection, Sexual Dimorphism in Home Range Sizes, and the Potential for Male Monopolization of Females

In both macropods and eutherians, male home ranges are largest whereas females are found on small home ranges or territories (Fisher and Owens 2000), a condition likely to induce sexually segregated and polygynous population architectures, the most common population structures in mammals. An inference from the previous authors' review is that territorial males whose ranges overlap those of females appear to represent an intermediate architecture. Deductions from the literature reviewed in the paper on macropods await quantitative testing, particularly, given the theoretical treatment by Rodrigues and Gardner (2013) showing that group-size effects are not straightforward and that temporal factors (e.g., climate) may have stronger effects than spatial factors (e.g., food dispersion) in many regimes (cf. macropods: Eisenberg 1981; cavies, Caviidae: Adrian and Sachser 2011).

The patterns specified by Rodrigues and Gardner (2013) have important implications for apparent differences between macropods and eutherians whose population parameters may be more affected by rainfall and ecological variables, respectively. Finally, the importance of variations in group size must be weighed by mammalian social biologists since small group size is ubiquitously assumed to correlate with higher grades of sociality in mammals and many other taxa (e.g., birds; Sect. 5.4; Synopsis). Rodrigues and Gardner's (2013) analysis questioned the latter assumption. However, their conclusions depended upon the relative viscosity (variations in dispersal distance) within populations, suggesting that some patterns of mammalian population structure may be a function of variations in dispersal rate (see Johnson and Gaines 1990; Waser et al. 2013).

Other caveats do obtain, however, since males' energetic requirements and intrasexual selection ("male–male competition"; this brief, Sect. 8.5) may yield male dispersions that are theoretically suboptimal for male reproductive benefits. In macropods, monogamy is associated with small, exclusive, female territories (Fisher and Owens 2000; also see Hennessy et al. 2012). Monogamy may also occur where female dispersion may be unpredictable or sparse, decreasing benefits or increasing

Table 3.2 Variations in spatiotemporal architecture of populations of caviomorph rodents (family Caviidae) relative to variations in environmental factors. (Based on Adrian and Sachser 2011)

Species	Spatiotemporal patiotemporal structure	Environmental correlate(s)	Comments
Kerodon rupestris	Harem, resource-defense polygyny	Females aggregate around rock piles	Males monopolize females
Cavia magna	Solitary-promiscuous	Habitats flood intermittently, access to food and water changeable	Females spatiotemporally unpredictable, preventing male monopolization
Cavia aperea	Pair, harem, female-defense polygyny	Humid habitat, abundant food (evenly distributed?), high predation pressure	Females evenly distributed, high population densities, male monopolization of females, does not burrow, "cryptic predator-avoidance strategy precludes the evolution of large groups" (Adrian and Sachser 2011), small foraging groups may be antipredator strategy
Galea musteloides	Solitary, multimale-multifemale, promiscuous	Dry habitat, sparse vegetation at some sites (variable), habitat generalist, food abundant at some sites, primarily a grazer	Females larger than males, paedomorphosis in males decreases competition between males and females driven by desiccation risk (?), sometimes sympatric with *C. aperea*
Microcavia australis	"Female-centered multimale-multifemale groups"	Eats leaves and fruits in trees and shrubs, high predation pressure	High population densities, highly gregarious, sometimes sympatric with *G. musteloides*, climbing an adaptation to decrease interspecific competition (?), foraging in groups may be antipredator strategy
Microcavia niata	Group-living	High predation pressure	Burrowing, predator vigilance

Note that resource dispersion is one among several possible environmental factors (e.g., predator detection and defense, mate guarding, "social thermoregulation"). Adrian and Sachser (2011) pointed out that group living in cavies may be driven by costs associated with burrowing, features that vary with characteristics of soils. The latter observations do not imply that resource dispersion is unimportant in some conditions, but implies, rather, that different factors may be targets of different selection intensities. Importantly, Adrian and Sachser (2011) noted that "different strategies might have arisen to solve the same problem," highlighting the need to study different "routes" to group living allowed by the same or similar "toolkit" (promiscuity?)

costs of polygyny, in this case, male ranges overlapping the ranges of > 1 female. Variations in the factors associated with the aforementioned conditions may yield insights into differences in social organization between monogamous mammals (e.g., *Potorous longipes,* and polygynous, *P. tridactylus*). Monogamy is rare in the primitive group, cavies (Table 3.2), suggesting that this sociosexual structure is highly derived (see Adrian and Sachser 2011).

In the mammalian literature, terminology for mating systems is often not differentiated from terminology for social systems. As well, use of terminology is
sometimes inconsistent. Terminology in mammalian social biology is strongly influenced by terminology found in the literature on birds with an emphasis on mating
systems ("polyandry," "leks," "monogamy," "polygyny," "polygynandry"). Perhaps,
the best example is use of the word, "polygyny" (see cavies, *Cavia aperea:* Rémy
et al. 2013; Asher et al. 2004), a type of population architecture whereby (1) a single male's home range or territory overlaps that of 1 or > 1 female home range or
territory, or (2) a single male in residence with 1 or > 1 female on a home range or territory. "Polyandry," also, is in need of clarification (cf. Andersson 1994: "classical"
polyandry and multiple mating by females). Despite "fuzziness" and lack of scientific consensus about terminology, the defining feature of mammalian population
structure is spatial dispersion rather than patterns of mating per se. The prior perspective strongly suggests that spatial, or spatiotemporal, dispersion "drives" coevolution
between sexual and social systems (Crespi 2007: Sect. 1.1), and that aggregations,
tolerance, and, possibly, group life are prior (see polygynous bank voles, *Myodes
glareolus:* Rémy et al. 2013).

Fisher and Owens (2000) provided a convincing analysis of the relationship between "mating systems and sex differences in home range size," and their schemas
correspond well with reviews and empirical reports on other terrestrial mammals
(e.g., platypus, *Ornithorhynchus anatinus:* Grant and Temple-Smith 1998; shrews,
Soricidae: Churchfield 1990; tree-shrews, Tupai: Emmons 2000; humans, *Homo
sapiens:* Lee and DeVore 1976; Meggitt 1965; short-tailed opossum, *Monodelphis
domestica:* Caramaschi et al. 2011; brown bears, *Ursus arctos:* Steyært et al. 2013;
ruminants, Conradt 1998, c.f.; Metatheria: Lee and Cockburn 1985). Despite the apparent generalities of the schemas presented by Fisher and Owens (2000), anomalies
remain (e.g., red acouchies, *Myoprocta exilis:* Dubost 1988; dasyurids, *Antechinus:*
Lee and Cockburn 1985; pentail tree-shrews, *Ptilocercus:* Emmons 2000). A possible
consideration is that temporal factors may be particularly constraining for these and
other anomalous taxa, whereas, as reviewed above and below, spatial factors appear to
explain many variations in mammalian population dispersions. Temporal and spatial
factors need to be decomposed quantitatively, including, experimentally, projects that
would reveal both strengths and weaknesses of spatiotemporal analyses (Chap. 6).

3.4 "Promiscuous" Associations with Overlapping Home Ranges Without Male Monopolization: A Mammalian "Toolkit"

Most eutherians demonstrate one or another type of "polygyny" (Eisenberg 1966,
1981; Wilson 1975), but macropod "social" organization is conserved and relatively
invariant, with nonterritorial, "promiscuous" structures being most common (Fisher
and Owens 2000; see Cornwallis et al. 2010 for a general model). In particular, the
home ranges of male and female macropods overlap one another, with little evidence

of either sex monopolizing the other. Supporting Eisenberg's (1981) proposition, it seems reasonable to suggest that the "promiscuous" dispersion is an evolved template providing a flexible "toolkit" for the elaboration of population structure in response to spatial and temporal, abiotic and biotic, including, intraspecific ($l^*_{between}$, l^*_{within}), effects. It follows from the discussion so far that short-term and long-term variations in "patch" and "population" density are expected to have been critical determinants of the differential reproductive costs and benefits to individuals of exhibiting mechanisms to manage competition. Although the present monograph is not intended to "unpack" the phylogenetic progression of mammalian social evolution, it seems likely that the "promiscuous" arrangements described for macropods constitute the primitive structures in the class, dependent upon tolerance possibly imposed environmentally by high population densities or, simply, by chance encounters of conspecifics during movements in "patches" and habitats (e.g., for mate search). Tolerance might have been particularly beneficial during dispersal or migration as well as any activity requiring search strategies.

By manipulating variables hypothesized to determine female home range or territory sizes, it seems to require uncomplicated evolutionary trajectories ("fast" or "slow": see Selman et al. 2012) from "promiscuous" population structure to any other of the social structures described for mammals. In addition to the critical factors and correlations already discussed in this chapter, it is useful to begin with the assumption that "female mammals are expected to minimize home ranges enabling them to forage widely enough to find sufficient food with minimum risk and energy expenditure" (Fisher and Owens 2000; McNab 1980). Depending, then, upon the condition-dependent "potential" for females to "maximize" reproductive success (via direct and/or indirect reproduction), males, time-minimizers, are expected to "maximize" the number of females monopolizeable. It is apparent that in mammals, males generally avoid coresidence with females. Of course, types, ultimately, do the best they can (Waser et al. 2013; Austad 1984), and, sometimes, males will do best by minimizing competition via tolerating or facilitating other reproductive males and/or females, such as, by forming multimale groups in association with females, a social structure that is relatively common in primates and a few other mammalian taxa (Fig. 2.1, Chap. 4, Sect. 8.5), virtually absent in Aves.

The flexibility of the "promiscuous" template echoes Lee's (1976) terminology for human social evolution, accordion-like "concentration and dispersion," a useful paradigm for mammals in general. Cavies, as noted, would make a good model system for investigations of the evolution of a range of grouping structures (this brief Table 3.2; see Rood 1972), consistent with the principles reviewed above as well as with spatiotemporal theories of group formation and group maintenance (Chap. 6). The reviews of marsupials by Lee and Cockburn (1985) and of cavies by Adrian and Sachser (2011) are consistent with the rules of population assembly reviewed by Fisher and Owens (Fisher and Owens 2000; see Andersson 2005 for a compatible analysis based on the "operational sex ratio"). It is clear from Lee and Cockburn's paper (2000) how the evolution of greater flexibility in social organization might be advantageous in heterogeneous regimes, as well as, how the evolution of monogamy (also see "temporary monogamy" in coyotes, *Canis latrans:* Gilbert-Norton et al.

2013) in habitats with an even food distribution is a good "fit" to the dispersion of animals when food is sparsely distributed.

An interesting aside is that most mammalian taxa demonstrate relative flexibility of population structure, even when the same basic architecture is retained across species in a family, such as, tree shrew (Tupaiidae) territoriality combined with solitary foraging differentially responsive to variations in population density (compare *Tupaia glis* and *T. longipes:* Emmons 2000). Cavies have become a model taxon for the flexibility and variability of mammalian grouping patterns (Table 3.2), deserving targeted programs of investigation from biochemical to higher levels of organization (Adrian and Sachser 2011; Asher et al. 2004; Meserve et al. 1984), particularly, in combination with their Old World relatives (e.g., Bathyergidae). Fleming et al. (1987) concluded that neotropical forests demonstrated less heterogeneity than Paleotropical forests, two dynamic states that might have predictive value when comparing and contrasting grouping patterns between the two regions.

Fisher and Owens' (2000, pp 1090–1091) discussion of the roles that males play in determining population architecture leads one to the conclusion that the intensity of male–male competition or the spatiotemporal unpredictability of females may stress males' time budgets (Sect. 8.6), leading them to adopt strategies that are not, theoretically, optimal (e.g., "monogamy," leks). Indeed, lekking, whereby males gather on an exclusive breeding ground visited by females who "choose" one or more displaying males, is rare in mammals compared to birds; although a few taxa exhibit elements of the social structure whereby females float (?) or search (?) male home ranges or territories during a breeding season or as a matter of course (e.g., some pinnipeds, sea otters, some ungulates). Importantly, lek and some other systems (Hemelrijk 1999; polygynandrous mantled howler monkeys, *Alouatta palliata:* Jones and Cortés-Ortiz 1998) exhibit "female emancipation" (Andersson 2005; Emlen and Oring 1977), and, as suggested above, some male strategies may have evolved in response to costs incurred from multiple mating by females (in the previous cases, male–male tolerance at stationary breeding areas; multiple male coresidence in bisexual groups on home ranges whereby males may exhibit tolerance or various types of sociality such as group defense or coalition formation), and male tolerance of multiple mating by females (Chap. 8). Each of these strategies may be expected to reduce costs of male–male competition, favoring superior males within ($l*_{within}$) or between ($l*_{between}$) groups via reductions in thermal stress and increased energetic efficiency (see Gittleman and Thompson 1988).

3.5 "Solitary" Mammals and Sexual Segregation Grade to Polygyny

What are the "drivers" of "solitary" ("sexually segregated") population structures in mammals? Factors intrinsic and extrinsic to populations are deterministic, particularly, the potential for males to monopolize females and the dispersion of resources required by females to reproduce. Table 3.2, for example, displays the fundamental relationship between promiscuity and sexual segregation in *C. magna ,* a transition probably dependent upon variations in population density whereby the ancestral, promiscuous state, responds to increasing density and increased competition among males for access to variably dispersed females (see Steyært et al. 2013). Let us assume that mammal populations are inherently responsive to environmental perturbations because of their evolution in heterogeneous regimes favoring traits characteristic of invasive taxa designed for rapid expansion into new regimes. An integral element of mammalian flexibility would be the ability to adjust responses to abiotic (e.g., soil gradients, breeding sites) and biotic (e.g., food dispersion, nutrient gradients, predation) changes across space (e.g., across patches or habitats) and time (e.g., across seasons), and factors opposing the evolution of sociality in the class (Jones 2009).

Compare *C. magna* and *G. musteloides* in Table 3.2 whereby, hypothetically, in the latter species, some threshold level of population increase induced deleterious effects on male reproductive success, increasing benefits of male–male tolerance (multimale–multifemale groups) and facilitation of females ("female dominance") in some regimes. In these cases, components of a "promiscuous" spatiotemporal structure without male monopolization of females and with overlapping male and female home ranges acts as a multipurpose ©Lego kit bounded by the environmental potential to accommodate reproductive tactics and strategies of types. Reviewing the literature on cavies, Adrian and Sachser (2011) stated: "Female behavior is obviously a decisive factor that prevents monopolization by males." Importantly, the latter authors' treatment highlights sexual conflict between mammalian males and females (Aloise King et al. 2013), as well as the coevolution of tactics and strategies, leading to the conclusion that population structure in cavies, and possibly other mammals, is an ultimate function of the dispersion and sizes of female home ranges relative to male thermal zones.

References

Adrian O, Sachser N (2011) Diversity of social and mating systems in cavies: a review. J Mammal 91:39–63

Alexander RD, Noonan KM, Crespi BJ (1991) The evolution of eusociality. In: Sherman PW, Jarvis JUM, Alexander RD (eds) The biology of the naked mole-rat. Princeton University Press, Princeton

Aloise King ED, Banks PB, Brooks RC (2013) Sexual conflict in mammals: consequences for mating systems and life history. Mamm Rev 43:47–58

Andersson M (1994) Sexual selection. Princeton University Press, Princeton

Andersson M (2005) Evolution of classical polyandry: three steps to female emancipation. Ethology 111:1–23

Asher M, Lippmann T, Epplen JT, Kraus C, Trillmich F, Sachser N (2004) Large males dominate: ecology, social organization, and mating system of wild cavies, the ancestors of the guinea pig. Behav Ecol Sociobiol 62:1509–1521

Austad SN (1984) A classification of alternative reproductive behaviors and methods for field-testing ESS models. Am Zool 24:309–319

Bodmer RE (1990) Ungulate frugivores and the browser-grazer continuum. Oikos 57:319–325

Bourke AFG (2011) Principles of social evolution. Oxford University Press, Oxford

Brook BW, Bowman DMJS (2002) Explaining the Pleistocene megafaunal extinctions: models, chronologies, and assumptions. Proc Nat Acad Sci 99:14624–14627

Caramaschi FP, Nascimento FF, Cerqueira R, Bonvicino CR (2011) Genetic diversity of wild populations of the grey short-tailed opossum, *Monodelphis domestica* (Didelphimorphi: Didelphidae), in Brazilian landscapes. Biol J Linn Soc 104: 251–263

Chapman JA, Feldhamer GA (eds) (1982) Wild mammals of North America: biology, management, and economics. Johns Hopkins University Press, Baltimore

Churchfield S (1990) The natural history of shrews. Cornell University Press, Ithaca

Conradt L (1998) Could asynchrony in activity between the sexes cause intersexual social segregation in ruminants? Proc Roy Soc Lond B 265:1359–1363

Cornwallis CK, West SA, Davis KE, Griffin AS (2010) Promiscuity and the evolutionary transition to complex societies. Nature 466:969–972

Crespi BJ (2007) Comparative evolutionary ecology of social and sexual systems: waterbreathing insects come of age. In: Duffy JE, Thiel M (eds) Evolutionary ecology of social and sexual systems: crustaceans as model organisms. Oxford University Press, Oxford

Di Stefano J, Coulson G, Greenfield A, Swan M (2011) Resource heterogeneity influences home range area in the swamp wallaby *Wallabia bicolor*. Ecography 34:469–479

Dubost G (1988) Ecology and social life of the red acouchy, *Myoprocta exilis*; comparison with the orange-rumped agouti, *Dasyprocta leporina*. J Zool 214:107–123

Eisenberg JF (1966) The social organizations of mammals. Handbüch der Zoologie 8:1–92

Eisenberg JF (1981) The mammalian radiations: an analysis of trends in evolution, adaptation, and behavior. University of Chicago Press, Chicago

Emlen ST, Oring LW (1977) Ecology, sexual selection, and the evolution of mating systems. Science 197:215–223

Emmons LH (2000) Tupai: a field study of Bornean treeshrews. University of California Press, Berkeley

Ewer RF (1968) Ethology of mammals. Logos, London

Fisher DO, Owens IPF (2000) Female home range size and the evolution of social organization in macropod marsupials. J Anim Ecol 69:1083–1098

Fleming TH, Breitwisch R, Whitesides GH (1987) Patterns of tropical vertebrate frugivore diversity. Ann Rev Ecol Syst 18:91–109

Gellner G, McCann K (2012) Reconciling the omnivory-stability debate. Am Nat 179:22–37

Gilbert-Norton LB, Wilson RR, Shivik JA (2013) The effect of social hierarchy on captive coyote (*Canis latrans*) foraging behavior. Ethology 119:335–343

Gittleman JL, Thompson SD (1988) Energy allocation in mammalian reproduction. Am Zool 28:863–875

Grant TR, Temple-Smith PD (1998) Field biology of the platypus (*Ornithorhynchus anatinus*): historical and current perspectives. Phil Trans Roy Soc Lond B 353:1081–1091

Hamilton WD (1964) The genetical theory of social behavior. J Theor Biol 7:1–52

Hemelrijk CK (1999) An individual-oriented model of the emergence of despotic and egalitarian societies. Proc Roy Soc Lond B 266:361–369

Hennessy CA, Dubach J, Gehrt SD (2012) Long-term pair-bonding and genetic evidence for monogamy among urban coyotes (*Canis latrans*). J Mammal 93:732–742

Jarmon PJ, Southwell CJ (1986) Grouping associations and reproductive strategies in Eastern grey kangaroos. In: Rubenstein DI, Wrangham RW (eds) Ecological aspects of social evolution: birds and mammals. Princeton University Press, Princeton

Jetz W, Rubenstein DR (2011) Environmental uncertainty and the global biogeography of cooperative breeding in birds. Curr Biol 21:72–78

Johnson ML, Gaines MS (1990) Evolution of dispersal: theoretical models and empirical tests using birds and mammals. Ann Rev Ecol Syst 21:449–480

Jones CB (2005) Social parasitism in mammals with particular reference to Neotropical primates. Mastozoologíca Neotrop 12:19–35

Jones CB (2009) The effects of heterogeneous regimes on reproductive skew in eutherian mammals. In: Hager R, Jones CB (eds) Reproductive skew in vertebrates: proximate and ultimate causes. Cambridge University Press, Cambridge

Jones CB (2012) Robustness, plasticity, and evolvability in mammals: a thermal niche approach. Springer, New York

Jones CB, Cortés-Ortiz L (1998) Facultative polyandry in the howling monkey (*Alouatta palliata*): Carpenter was correct. Bol Primatol Lat 7:1–7

Kermack DM, Kermack KA (1984) The evolution of mammalian characters. Croom Helm, London

Ketola T, Kellermann VM, Loeschke V, López-Sepulcre A, Kristensen TN (2013) Does environmental robustness play a role in fluctuating environments? Evolution. doi:10.1111/evo.12285

Kielan-Jaworowska Z, Cifelli RL, Luo Z-X (eds) (2004) Mammals from the Age of Dinosaurs: evolution and structure. Columbia University Press, New York

Lacey EA (2000) Spatial and social systems of subterranean rodents. In: Lacey EA, Patton JL, Cameron GN (eds) Life underground: the biology of subterranean rodents. The University of Chicago Press, Chicago

Ladevèze S, de Muizon C, Beck RMD, Germain D, Cespedes-Paz R (2011) Earliest evidence of mammalian social behaviour in the basal Tertiary of Bolivia. Nature 474:83–86

Lee AK, Cockburn A (1985) Evolutionary ecology of marsupials. Cambridge University Press, Cambridge

Lee RB (1976) !Kung spatial organization: an ecological and historical perspective. In: Lee RB, DeVore I (eds) Kalahari hunter-gatherers: studies of the !Kung San and their neighbors. Harvard University Press, Cambridge

McNab BK (1980) Food habits, energetics, and the population biology of mammals. Am Nat 116:106–124

Matsuda I, Tuogo A, Bernard H, Sugau J, Hanya G (2013) Leaf selection by two Bornean colobine monkeys in relation to plant chemistry and abundance. Sci Rep 3. doi:10.1038/srep01873

Meggitt MJ (1965) Desert people: a study of the Walbiri aborigines of central Australia. University of Chicago Press, Chicago

Meserve PL, Martin RE, Rodriguez J (1984) Comparative ecology of the caviomorph rodent *Octodon dégus* in two Chilean Mediterranean-type communities. Revista Chiléna de Historia Natural 57:79–89

Milligan HT, Koricheva J (2013) Effects of tree species richness and composition on moose winter browsing damage and foraging selectivity: an experimental study. J Anim Ecol 82:739–748

Morales MM, Giannini NP (2013) Ecomorphology of the African felid ensemble: the role of the skull and postcranium in determining species segregation and assembling history. J Evol Biol. doi:10.1111/jeb.12108

Owen-Smith N, Chafota J (2012) Selective feeding by a megaherbivore, the African elephant (*Loxodonta africana*). J Mamm 93:698–705

Packer C (1986) The ecology of sociality in felids. In: Rubenstein DI, Wrangham RW (eds) Ecological aspects of social evolution: birds and mammals. Princeton University Press, Princeton

Rakotoarisoa JE, Raheriarisena M, Goodman SM (2013) Late Quaternary climatic vegetational shifts in an ecological transition zone of northern Madagascar: insights from genetic analyses of two endemic rodent species. J Evol Biol. doi:10.1111/jeb.12116

Rémy A, Odden M, Richard M, Tyr Stene M, Le Gaillard J-F, Andreassen HP (2013) Food distribution influences social organization and population growth in a small rodent. Behav Ecol. doi:10.1093/beheco/art029

Rodrigues AMM, Gardner A (2013) Evolution of helping and harming in viscous populations when group size varies. Am Nat 181:609–622

Rood JP (1972) Ecological and behavioral comparisons of three genera of argentine cavies. Anim Behav Monog 5:1–83

Roughgarden J (1979) Theory of population genetics and evolutionary ecology: an introduction. Macmillan, New York

Sedio BE, Ostling AM (2013) How specialized must natural enemies be to facilitate coexistence among plants? Ecol Lett. doi:10.1111/ele.12130

Selman C, Blount JD, Nussey DH, Speakman JR (2012) Oxidative damage, ageing, and life-history evolution: where now? Trends Ecol Evol 27:570–577

Smith FA, Boyer AG, Brown JH, Costa DP, Dayan T, Morgan Ernest SK, Evans AR, Fortelius M et al (2010) The evolution of maximum body size of terrestrial mammals. Science 330:1216–1219

Stearns SC, Koella JC (1986) The evolution of phenotypic plasticity in life-history traits: predictions of reaction norms for age and size at maturity. Evolution 40:893–913

Steyært SMJG, Kindberg J, Swenson JE, Zedrosser A (2013) Male reproductive strategy explains spatiotemporal segregation in brown bears. J Anim Ecol 82:836–845

Stahler DR, MacNulty DR, Wayne RK, vonHoldt B, Smith DW (2013) The adaptive value of morphological, behavioural and life-history traits in reproductive female wolves. J Anim Ecol 82:222–234

Trivers RL (1972) Parental investment and sexual selection. In: Campbell B (ed) Sexual selection and the descent of man. Aldine, New York, pp 1871–1971

Turner A (2004) Prehistoric mammals. National Geographic Press, Washington, DC

Waser PM, Nichols KM, Hadfield JD (2013) Fitness consequences of dispersal: is leaving home the best of a bad lot? Ecology 94:1287–1295

Wilson EO (1975) Sociobiology: the new synthesis. Belknap, Cambridge

Wilson GP, Evans AR, Corfe IJ, Smits PD, Fortelius M, Jernvall J (2012) Adaptive radiation of multituberculate mammals before the extinction of dinosaurs. Nature 483:457–460

Woodard SH, Fischman BJ, Venkat A, Hudson ME, Varala K, Cameron SA, Clark AG, Robinson GE (2011) Genes involved in convergent evolution of eusociality in bees. Proc Nat Acad Sci U S A 108:7472–7477

Chapter 4
Multimale-Multifemale Groups and "Nested" Architectures: Collaboration Among Mammalian Males

> *Multi-male troops are distinguished from the age-graded-male troops by the presence typically of two or more males who are full adults, physically and behaviorally.*
>
> *Brown (1975)*

Abstract This chapter addresses the evolution of multimale-multifemale groups and the evolution of "nested" societies. In the former, > 1 reproductive males co-reside with > 1 reproductive females as well as their young, and are generally characterized by multiple mating by females ("polyandry," "female emancipation"). Mechanisms such as dominance hierarchies and "queuing" manage competition in the foregoing reproductive units, and types of both sexes usually display tolerance, if not facilitation, among unrelated adults. Multimale-multifemale and "nested" population structures exhibit incipient division of labor, comparable, in some ways, to features associated with "primitively eusocial" mammals and social insects.

Keywords Incipient division-of-labor · Multimale-multifemale groups · Phenotypic diversity · Queuing · Fitness optima · Asymmetries

Feldhamer et al. (2004) proposed that mammalian sociality is characterized by the presence of alarm calling, cooperative rearing of young, coalitions and alliances, as well as eusociality; however, all of these features are not necessarily displayed in concert in a given taxon. Chapter 3 reviewed the variety of sociosexual structures in the Class, suggesting that mammalian groups form promiscuous aggregations of reproductive males and females with nonoverlapping or nonexclusive home ranges or territories. In this chapter the latter pattern continues. In some taxa, > 1 related or unrelated reproductive males demonstrate a significant degree of interindividual tolerance (Table 2.1, Table 3.2), located in aggregations varying in their spatiotemporal integration, coordination, and persistence. Multimale-multifemale and "nested" structures are noteworthy for the aforementioned diagnostic characters, and Wilson (1975) reported that, among mammals, multimale-multifemale or "nested" assemblages are found among some macropods, bats, sciurids, mice and dormice, chinchillas, whales, dolphins, porpoises, carnivores, pinnipeds, perissodactyls, artiodactyls, elephants, as well as, primates, including humans. The terminology for these structures has not been standardized; however, it is clear from Wilson's (1975) summary that most multimale-multifemale and "nested" assemblages represent

C. B. Jones, *The Evolution of Mammalian Sociality in an Ecological Perspective,*
SpringerBriefs in Ecology, DOI 10.1007/978-3-319-03931-2_4,
© Clara B. Jones 2014

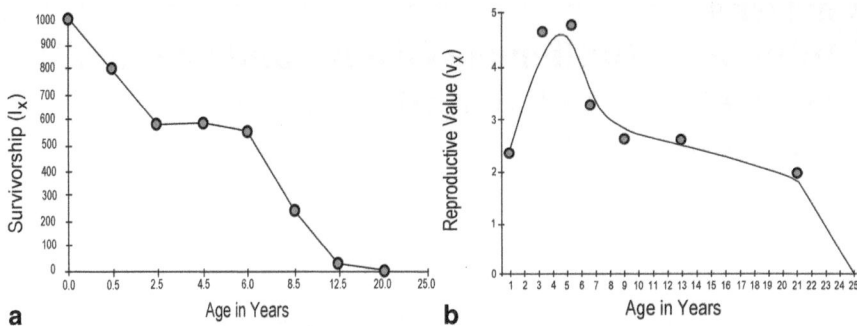

Fig. 4.1 v_x: the relative number of offspring produced by each female surviving to age x

aggregations of reproductives of both sex, sometimes exhibiting male dominance hierarchies, but without extensive integration and coordination of group members or differentiation into roles. The latter traits are especially evident among some Primates (e.g., *Macaca, Papio,* atelids, human populations); thus, most examples in this chapter concentrate on the latter order. Multimale-multifemale and "nested" primate societies include types exhibiting many "non-damaging" mechanisms to manage competition (e.g., coercive, reward, expert, legitimate, and referent "power": French and Raven 1959 in Jones 2000), topics in need of investigation since they suggest the evolution of high social "grades" (thus, "incipient" eusociality).

4.1 Incipient Division-of-Labor in Multimale-Multifemale Groups of Mammals

Incipient, facultative, or temporary division-of-labor may characterize a number of mammalian species exhibiting multimale-multifemale or "age-graded" group struc-tures (e.g., elephants, *Elephas* spp.: McComb et al. 2011; Robinson et al. 2012; African lions, *Panthera leo*: Packer et al. 2001; black howler monkeys, *Alouatta pigra*: Van Belle et al. 2013; bottlenose dolphins, *Tursiops*: Lusseau 2007; African striped mouse, *Rhabdomys pumilio*: Rymer and Pillay 2013). Based on a study of one aged, marked group of mantled howler monkeys (*Alouatta palliata* Gray) in riparian habitat of tropical dry forest in Costa Rica in 1976 and 1977, Jones (1996) reported temporal division-of-labor in the form of age-dependent foraging behavior, a trait similar to social bathyergids (see Lacey and Sherman 1997) and some social insects (Wilson 1971). Inferior competitors in a "patch" may benefit from facilitating the reproduction of conspecifics as a mechanism to manage conflict with superiors, in the mantled howler case, investing time and energy to engage in group foraging. Since increasing age or size eventually entails decreasing reproductive value (v_x: the relative number of offspring produced by each female surviving to age x, Fig. 4.1), under some conditions the expression of "helping" would be positively associated with age. The latter proposition was tested by calculating a monthly foraging rate for each adult female in the focal group (Methods in Jones 1996), where foraging

Table 4.1 Age class, estimated age in years, number of females in each age class (N), observed (O), and expected (E) frequencies of social foraging, and cumulative chi square (X^2) values for a test of the null hypothesis. (Details can be found in Jones (1996) and this review's text)

Age class (yrs)	N	O	E	$(O-E)^2/E$
Young adult (5–7)	5	15	42.4	17.71
Middle-aged (7–10)	5	35	42.4	1.29
Middle-aged to Old (10–15)	1	18	8.1	12.11
Old (15+)	1	33	8.1	76.54
$\sum$	12	101	101.0	107.65

was defined as a female leaving her group to locate a food source, giving a unique vocalization (Jones 1998) recruiting other group members to the source (hereafter, "social foraging"). The terminal source was identified 52 times, and most of these social foraging bouts (85 %, $n = 44$) resulted in feeding on ephemeral food, new leaves, flowers, or fruit (the monkeys' preferred food) rather than on mature leaves (15 %, $n = 8$).

   Table 4.1 presents results of my analysis of social foraging as a function of female age, including, expected frequencies, in one group of mantled howler monkeys in Costa Rican riparian habitat. Computing "goodness-of-fit" led to an unequivocal rejection of the null hypothesis ($P \leq 0.001$, $X^2 = 107.64$, $df = 3$). Thus, old age and social foraging frequency are significantly related. Young adult females initiate foraging significantly less than expected on the basis of their numbers ($P \leq 0.001$), suggesting that these females are "selfish," conserving time and energy for direct reproduction or for competition. Table 4.1 also shows that the middle-aged to old female engaged in social foraging more than expected by chance ($P \leq 0.01$), and this female succeeded the oldest and lowest ranking female as the most frequent forager when the old female disappeared from this group in 1977, following a nasty encounter with a prehensile-tailed porcupine (*Coendou*). A higher grade of temporal division-of-labor may be constrained in mantled howlers because of bisexual dispersal or other factors (e.g., environmental heterogeneity: Jones 2009; see Taylor et al. 2013) increasing geneflow or outbreeding, or because limiting resources are not sufficiently predictable to favor a social trajectory. The latter conditions will limit the benefits of facilitation among non-kin and, possibly, among kin. Except in the sense that dispersal from natal groups can be considered altruistic (Chap. 2), mantled howlers are not gregarious or obvious facilitators though, like other members of Atelidae, they exhibit relatively low rates of aggression when interacting with other group members compared to many other polygynandrous mammals, including, primate, taxa.

4.2 "Nested" ("Hierarchical," "Modular") Social Organization and Management of Competition as Well as Competition by Queuing

"Nested" ("hierarchical," "multi-level," "modular") effects are ubiquitous in nature since every type interacts with components of an environment (Fig. 1.3, Fig. 2.2). A unit at a lower level (scale) of organization is "nested" within units at higher scales, etc., and interactions may be one-way, two-way, or indirect (Schneider and Brosell 2012; Chap. 7, this book), incorporating, as well, feedbacks (e.g., plant–plant, plant–herbivore, herbivore–herbivore: Chap. 7). "Nested" designs are common to many social taxa, from social insects to humans, and, among social mammals, "nested" architectures characterize female elephant, Hamadryas baboon (*Papio hamadryas*), some cetacean group structures, and a few others (Grueter et al. 2012). The basic unit of a "nested" population architecture is either polygyny or polygynandry (Grueter et al. 2012), a proposition in accord with the "toolkit" concepts addressed throughout this review. It seems likely that each of the latter structures, discussed previously (Chap. 3), probably derived from a promiscuous baseline via different ecological and competitive regimes. If only because of the conformations of hierarchical societies, ample opportunities may exist for individuals, subgroups, and groups to be subordinated or exploited by other such units due to intrinsic (e.g., genetic, physiological, including developmental) and extrinsic (thermal zone, "patch") asymmetries, influencing l^*_{within} and $l^*_{between}$. In humans, the latter differences are associated with caste structures (Dumont 1981), "slavemaking" ("slavery": Davis 1984), and other roles based upon biologically or culturally inferior status (Lotka 1928; "incipient" division-of-labor?), leading Jones (2011) to suggest that *Homo sapiens* might be characterized as "facultatively eusocial."

Students of "nested" societies hold that these structures are "mutualistic assemblages" between levels of (whole organism) organization (e.g., Lusseau 2007; Lusseau et al. 2006; Hoelzel et al. 2007; Robinson et al. 2012). However, to my knowledge, this assumption has not been tested quantitatively, including mathematical modeling or experiments. Based on the evidence provided in a 2012 International Journal of Primatology special issue (no. 33) on "multilevel societies in primates and other mammals," it does appear that interindividual and intergroup tolerance is a defining feature of these structures (see Grueter et al. 2012; Fig. 6). It also seems likely that "nested" architecture may function as "information centres" increasing efficient search and exploitation of resources, decreasing l^* values for some types. The latter population structure requires study in a wider range of mammalian orders, particularly, bats (see Wilkinson 1992).

The flexibility of "nested" human structures is associated with the breathtaking capacity for exploitation of global biomes by this species, seemingly demonstrating the effectiveness of the multilevel associations for coordination and control of group members, impacting values of l^* for each type. Following Jones (2013; see this brief, Sect. 3.4), though many aspects of human biology are relatively well known, the capacity of ancient and modern technological societies to maintain high population densities (high α-diversity), to successfully invade virtually all global habitats (high β-diversity), to modify their areal ranges (high γ-diversity), to utilize effective mechanisms of niche invasion and expansion (e.g., collaboration, social learning, fire, tools, migration, war), and to impose profound, deleterious effects on biogeochemistry demand systematic treatments of hominin ecology, phylogeny, and evolution, including comparisons and contrasts with other "invasive species."

Frank's (2013) treatment suggests that phenotypic diversity will be induced by novel (e.g., disappearance of a limiting resource) or extreme (e.g., severe drought)

environmental events and that responses may be genetic (mutation), cellular (physiological and developmental), or learned (e.g., by trial-and-error, by association).
Applied to humans, Frank's (2013) treatment suggests that human characteristics
such as collaboration, tool use, the application of fire for processing food, the manu-
facture of clothing, language, long-distance dispersal, social learning, higher-order
problem solving, and the like, effectively switch an environment ("landscape") from a
stressful (difficult, dangerous, risky, extreme, novel), "rugged" one initially decreasing reproductive rates (see Sibley and Hone 2003), to a less stressful, more even,
or "smoother" one (increasing reproductive rates and lowering l^* for a given type).
Fluctuating environments, including stressful interactions with other humans within
and between populations, will negatively impact l^* values (increasing them), increasing intensities of competition that, in humans, appear to have resulted in a variety of
tactics and strategies for coexistence (facilitation, tolerance, and/or inhibition). Flexible phenotypes, it is proposed, connect or extend one type or group of human to one
or more resource patches, including, other human types and groups, thereby, broadening the effective space of phenotypes, decreasing deleterious consequences of
environmental challenges, mediating l^*_{within} and $l^*_{between}$. Types competing within-
or between-groups scramble for or contest resources to yield effective reproductive
(type) or growth (group or population) rates, attempting to avoid lethal or sublethal
conditions via differential life history tactics and strategies (Table 1.1). l^* values can
be expressed as "genotypic specific" viabilities across thermal regimes that, for most
mammals, will be characterized by varying patterns of environmental fluctuation
(see Ketola et al. 2013).

A feature differentiating human "nested" societies from those of other mammal groups with similar social organization may be the highly refined ability of
humans to utilize damaging (coercion, force, social parasitism) and non-damaging responses, including tolerance, collaboration, facilitation, inhibition, and exploitation
(Table 2.1), to conditionally manage competition ("repression of competition": Frank
1995, 2003), enhancing opportunities for despotism (dominance, high skew) by superior types (reflecting l^*_{within} and $l^*_{between}$). As pointed out by Jones (2013), populous
human societies, and other "nested" assemblages, may maintain spatiotemporally
predictable queues at different scales of organization based upon interindividual
asymmetries governed by shared rules transmitted by social learning (see Krützen
et al. 2005). Some human populations appear more flexible (USA, UK, France,
Brazil) than others (China, Japan, North Korea, Scandinavia), features that may
be explained by variations in conditionally expressed traits associated with reaction norms and facultative division-of-labor (Jones 2011). The modular nature of
human "nested" architectures is well exemplified by patterns of networks inherent
to colonial societies with a central authority from which units of varying influence
extend. Patterns of "concentration and dispersion" (Lee 1976; Tanaka 1973) could
be mapped throughout human history, including, the role(s) of "slavemaking" during
the expansion phases of empires and colonial states, an operation similar to that observed for social insect "slavemaking" societies (CB Jones, unpublished data). In the
social insect literature, "slavemaking" is classified as a form of "social parasitism,"
a particular type of exploitation (Jones 2007).

Following Frank's (2013) conceptual framework, Jones (2013) posited that numerous traits characterizing *Homo sapiens* served to decrease environmental challenges deleterious to reproductive rates (e.g., abundance of limiting resources, low environmental fluctuation). The previous scenario might decrease asymmetries between the l^* values of some types, having an equalizing effect. Tanaka's (1976) studies of the $\neq$Kade San ("bushmen"), hunter-gatherers in the Kalahari (southern African desert), clearly demonstrated ways in which a cultural innovation limits mortality and, by extension, enhances reproductive success. The $\neq$Kade San, comprising mobile and mobile-subsistence units, inhabit a "marginal" environment characterized by drought (Tanaka 1976, Fig. 4.1, p. 105) and seasonal patterns of food availability (Tanaka 1976, Fig. 4.2, p. 108), a spatiotemporal regime not unlike the heterogeneous environments in which humans are thought to have evolved (Hill et al. 2011). On one occasion, Tanaka (1976) observed chacma baboons (*Papio ursinus*) foraging in the Kalahari, noting that this monkeys' home range was limited by their inability to cross arid land. The latter researcher compared the monkeys' habits with those of the $\neq$Kade San, capable of inhabiting the extreme desert environment as a result of digging through soil surface to locate and utilize the limiting resource, water. This cultural practice permitted a human "band" to expand inherent capacities, a "smoothing" effect decreasing likelihoods of sublethal or lethal outcomes, and increasing the likelihood of contacts with other "bands." Such phenotypic diversity is expected to impact individual life histories (survival, reproduction, growth), enhancing mean fitness of populations via increased reproductive rates, with consequent effects on higher levels of ecological organization (communities, ecosystems, biomes).

4.3   Queuing May Minimize *l** Levels for Superior Types

Queuing permits sorting and large group size where dominants or their representatives manage interindividual competition for resources convertible, directly or indirectly, to offspring quantity or quality by persuasion, coercion, force, or exploitation, in addition to positive and negative reinforcement (see synopsis). In human cultures, social rules transmitted via social learning activate repression of selfishness (see Frank 1995, 2003), achieving varying degrees of tolerance, collaboration, and/or cooperation. In large human societies, conflicts may erupt among relatives and nonrelatives via the breakdown of mechanisms to manage competition. However, the latter costs may be outweighed, on average, by benefits from efficient production of synergistic coordination and control. Furthermore, queuing systems avoid some costs attendant to division-or-labor in social insect societies by permitting relative flexibility and differentiation of roles maintaining dynamic networks of stable or unstable competition. Systematic, quantitative investigations of the stability (over time and space) of groups with large social networks have not received sufficient attention.

Jones' (2013) (applying Frank 2013) treatment of the ways in which phenotypic diversity and phenotypic novelty serve individual interests by facilitating lifetime reproductive success provides a schema that can be applied to most human tactics and strategies. In particular, the model permits researchers to evaluate the extent to which human responses to dynamic environmental challenges promote intentional problem-solving in extreme conditions ("cognitive" strategies), leading to mechanisms to manage competition within- and between-groups, influencing the inferior or superior status of types (i.e., influencing l^*_{within} and $l^*_{between}$). Interspecific competition may have been managed via mechanisms of domestication, artificial breeding, and extirpation in the service of superior humans, decreasing R^* for this species. As Jones (2013) pointed out, humans appear to combine a high fertility rate, high "reproductive effort," and long lifespan, like social Bathyergids, comparisons worthy of investigation. This combination of traits is not usually associated with mammals in heterogeneous ("rugged") regimes (Millar and Zammuto 1983). Similarly, most mammals are poor colonizers, and social mammals are generally constrained by their dependence upon conspecifics and group life (Cody 1986), challenges that humans have overcome to some degree via the "concentration and dispersion" spatiotemporal patterns and multilevel societies described by Lee (1976), Tanaka (1976), and Yellin (1976), combined with sophisticated, enforced, mechanisms of coexistence, and other tactics and strategies, including, cultural operations.

Differential reproductive costs and benefits of genotype x environment interactions require systematic investigation, including Bourke's (2011) assumption that stability only arises in associations between kin or between types sharing reproductive "fates." In "hierarchical" and other complex societies, problems associated with temporal and spatial coordination and control must be managed (e.g., Alberts et al. 2003), and the theoretical literature on "scheduling" indicates that many of these challenges are solved via within- and between-group "queuing" (Andrews et al. 2004), sorting processes with the potential to "buy" humans and similar mammalian species a degree of organizational flexibility typically associated with caste-forming social insects; although it is expected that flexibility will trade-off with efficiency, enhancing conflicts of interindividual interest and, possibly, aggressive interactions. Control of scheduling and sorting mechanisms may provide opportunities for some superior types, alone or in concert, to dominate and exploit others in the poorest conditions (low l^*_{within} and low $l^*_{between}$), managing and controlling limited positions in a queue. Similar to Chesson's (2000; see also Amarasekare 2003) logic and evidence from studies of community ecology (e.g., Kinahan and Pillay 2008), the higher likelihood of finding similar types within groups compared to between groups, are expected to yield higher intensities of competition within groups, in some conditions, favoring the evolution of mechanisms to manage competition.

References

Alberts SC, Watts HE, Altmann J (2003) Queuing and queue-jumping: long-term patterns of reproductive skew in male savannah baboons, *Papio cynocephalus*. An Behav 65:821–840

Amarasekare P (2003) Competitive coesistence in spatially structured environments: a synthesis. Ecol Lett 6:1109–1122

Andrews M, Kumaran K, Ramanan K, Stolyar A, Vijayakumar R, Whiting R (2004) Scheduling in a cueing system in asynchronously varying service rates. Prob Eng Info Serv 18:191–217

Bourke AFG (2011) Principles of social evolution. Oxford University Press, Oxford

Brown JL (1975) The evolution of behavior. W. W. Norton & Co., New York

Chesson P (2000) Mechanisms of maintenance of species diversity. Ann Rev Ecol Syst 31:343–366

Cody ML (1986) Diversity, rarity, and conservation in Mediterranean climate regions. In: Soulé ME (ed) Conservation biology. Sinauer, Sunderland, pp 123–152

Davis DB (1984) Slavery and human progress. Oxford University Press, New York

Dumont L (1981) *Homo hierarchicus*: caste system and its implications. The University of Chicago Press, Chicago

Feldhamer GA, Drickamer LC, Vessey SH, Merritt JF (2004) Mammalogy: adaptation, diversity, ecology, 2nd edn. McGraw-Hill, Boston

Frank SA (1995) Mutual policing and repression of competition in the evolution of cooperative groups. Nature 377:520–522

Frank SA (2003) Repression of competition and the evolution of cooperation. Evolution 57:693–705

Frank SA (2013) Natural selection. II. Developmental variability and evolutionary rate. J Evol Biol 24:2310–2320

Grueter CC, Li D, Ren B, Li M (2012) Overwintering strategy of Yunnan snub-nosed monkeys: adjustments in activity scheduling and foraging patterns. Primates. doi:10.1007/s10329-12-0333-3

Hill KR, Walker RS, Božičević M, Eder J, Headland T, Hewlett B, Hurtado AM, Marlowe F, Wiessner P, Wood B (2011) Co-residence patterns in hunter-gatherer societies show unique human social structure. Science 331:1286–1289

Hoelzel AR, Hey J, Dahlheim MS, Nicholson C, Burkanov V, Black N (2007) Evolution of population structure in a highly social top predator, the killer whale. Mol Biol Evol 24:1407–1415

Jones CB (1996) Temporal division of labor in a primate: age-dependent foraging behavior. Neotrop Primates 4:50–53

Jones CB (1998) A broad-band contact call by female mantled howler monkeys: implications for heterogeneous conditions. Neotrop Primates 6:38–40

Jones CB (2000) *Alouatta palliata* politics: empirical and theoretical aspects of power. Primate Report 56:3–21

Jones CB (2007) The evolution of exploitation in humans "Surrounded by strangers I thought were my friends." Ethology 113:499–510

Jones CB (2009) The effects of heterogeneous regimes on reproductive skew in eutherian mammals. In: Hager R, Jones CB (eds) Reproductive skew in vertebrates: proximate and ultimate causes. Cambridge University Press, Cambridge

Jones CB (2011) Are humans cooperative breeders? Arch Sex Behav 40:479–481

Jones CB (2013) Constraints on speciation in human populations: phenotypic diversity matters. Hum Biol Rev 2:263–279

Ketola T, Kellermann VM, Loeschke V, López-Sepulcre A, Kristensen TN (2013) Does environmental robustness play a role in fluctuating environments? Evolution. doi:10.1111/evo.12285

Kinahan AA, Pillay N (2008) Does differential exploitation of folivory promote coexistence in an African savannah granivorous rodent community? J Mammal 89:132–137

Krützen M, Mann J, Heithaus MR, Connor RC, Bejder L, Sherwin WB (2005) Cultural transmission of tool use in bottlenose dolphins. Proc Nat Acad Scis USA 102:8939–8943

Lacey EA, Sherman PW (1997) Cooperative breeding in naked mole rats: implications for vertebrate and invertebrate sociality. In: Solomon NG, French JA (eds) Cooperative breeding in mammals. Cambridge University Press, New York

Lee RB (1976) !Kung spatial organization: an ecological and historical perspective. In: Lee RB, DeVore I (eds) Kalahari hunter-gatherers: studies of the !Kung San and their neighbors. Harvard University Press, Cambridge, pp 73–97

Lotka AJ (1928) Sterility in American marriages. Proc Nat Acad Sci USA 14:99–109

Lusseau D (2007) Evidence for social role in a Dolphin social network. Evol Ecol 21:357–366

Lusseau D, Wilson B, Hammond PS, Grellier K, Durban JW, Parsons KM, Barton TR, Thompson PM (2006) Quantifying the influence of sociality on population structure in bottlenose dolphins. J An Ecol 75:14–24

McComb K., Shannon G., Durant SM, Sayialel K, Slotwo R, Poole J, Moss C (2011) Leadership in elephants: the adaptive value of age. Proc Roy Soc Lond B. doi:10.1098/rspb.2611.0168

Millar JS, Zammuto RM (1983) Life histories of mammals: an analysis of life tables. Ecology 64: 631–635

Packer C, Pusey AE, Eberly LE (2001) Egalitarianism in female African lions. Science 293: 690–693

Robinson MR, Mar KU, Lummaa V (2012) Senescence and age-specific trade-offs between reproduction and survival in female Asian elephants. Ecol Lett. doi:10:1111/j.1461-0248-2011.01735.x

Rymer TL, Pillay N (2013) Alloparental care in the African striped mouse Rhabdomys pumilio is age-dependent and influences the development of paternal care. Ethology. doi:10.1111/eth.12175

Schneider FD, Brosell U (2012) Beyond diversity: how nested predator effects control ecosystem functions. J An Ecol. doi:10.1111/1365-2656.12010

Sibley RM, Hone J (2003) Population growth rate and its determinants: an overview. In: Sibley RM, Hone J, Clutton-Brock TH (eds) Wildlife population growth rates. Cambridge University Press, Cambridge

Tanaka J (1973) Social structure of the Bushmen. In: Carpenter CR (ed) Behavioral regulators of behavior in primates. Bucknell University Press, Lewisburg

Tanaka J (1976) Subsistence ecology of Central Kalahari San. In: Lee RB, DeVore I (eds) Kalahari hunter-gatherers. Harvard University Press, Cambridge

Taylor TB, Rodrigues AMM, Gardner A, Buckling A (2013) The social evolution of dispersal with public goods cooperation. J Evol Biol. doi:10:1111/jeb.12259

Van Belle S, Estrada A, Garber PA (2013) Collective group movement and leadership in wild black howler monkeys (*Alouatta pigra*). Behav Ecol Sociobiol 67: 31–41

West, S.A., Pen, I. & Griffin, A.S. (2002). Cooperation and competition between relatives. Science: 296: 72–75

Wilkinson GS (1992) Information transfer at evening bat colonies. Anim Behav 44: 501–518

Wilson EO (1971) The insect societies. Belknap, Cambridge

Wilson EO (1975) Sociobiology: the new synthesis. Belknap, Cambridge

Yellin JE (1976) Settlement patterns of the !Kung: an archaeological perspective. In: Lee RB, DeVore I (eds) Kalahari hunter-gatherers. Harvard University Press, Cambridge

Chapter 5
Higher "Grades" of Sociality in Class Mammalia: Primitive Eusociality

> *Hamilton's rule forbids the evolution of altruism when relatedness is zero, regardless of the levels of benefit and cost. The evolution of altruism among non-relatives is conspicuously absent.*
>
> (Bourke 2011)

Abstract This chapter addresses "primitive eusociality" among mammals, including "cooperatively-breeding" species and the social mole rats (Bathyergidae). Like some social insects, "primitively eusocial" mammals are characterized by overlap of generations, cooperative breeding, and reproductive division-of-labor, though their phenotypes remain totipotent (capable of performing most tasks). Consistent with findings in other eusocial taxa, social mole rats utilize abundant, evenly dispersed nutrients and reside in a protective refugium. Interpretations of the literature presented in this chapter reinforce this brief's extension of the competitive coexistence literature in Community Ecology. Types' states are viewed herein as features evolved in response to competitive regimes rather than as deterministic functions of resource dispersion, per se.

Keywords "Routes" to sociality · Cooperative breeding · Primitive eusociality · Environmental heterogeneity · Constraints on social evolution

"Cooperatively-breeding" mammals and social bathyergids are eusocial, exhibiting overlapping generations, cooperative breeding, and reproductive division-of-labor. Because "reproductive skew" (relative apportionment of reproduction within groups) is steeper and some ecological features differ between eusocial mole rats and cooperative breeders (Alexander et al. 1991, Sherman et al. 1995; this brief, Preface), there is justification for retaining distinctive terminology, obfuscation that will be resolved by future studies. Neither cooperatively breeding mammals nor social bathyergids display sterile castes, retaining reproductive totipotency. Thus, cooperatively breeding mammals and social mole rats are a "primitive" form compared to many social insects, and both taxa are referred to herein as "primitively eusocial" mammals. Alexander et al. (1991) held that, besides social mole rats, pack-living mammals (some canids) should be considered eusocial. As many authors have pointed out (Crespi and Choe 1997), definitions of "eusociality" are often confusing, calling for consensus among social biologists.

5.1 Primitively Eusocial "Cooperative" Breeders

The edited volume by Solomon and French (1997) remains the classic reference on "cooperatively breeding" mammals. In cooperatively breeding societies, one or a few dominant females coexist with non-breeding "helpers," usually daughters or other female kin (Saltzman 2003). Primitively eusocial and advanced eusocial (some social insects and a few other animals: see Wilson 1971) architectures are presumed to represent energy-saving assemblies in heterogeneous environments (Russell et al. 2003), a hypothesis that has rarely been subjected to empirical tests for mammals. Studying 267 bird species, Cornwallis et al. (2010, 2009) investigated the relationship between promiscuity and the evolution of cooperative breeding, a group structure thought to have arisen from monogamy (see this brief, Fig. 3.2). Using a phylogenetic analysis, the previous authors tested the hypothesis that "the evolution of cooperative behavior is favored by low levels of promiscuity, leading to high within-group relatedness." Cornwallis et al. (2010) found that, while promiscuity was usually associated with reduced incidence of cooperative breeding in birds, "there are many promiscuous, cooperative species." The previous study demonstrated theoretically that "when promiscuity rates are very high or very low, variance in relatedness between broods will be low," leading to the prediction that selection on helping based on kin discrimination (altruism directed non-randomly by genotype toward group members) would be low at the extremes of the inverted-U distribution. However, at intermediate levels of promiscuity, cooperative breeding may evolve where potential helpers and donors have the capacity to discriminate kin. Subsequently, the previous authors' empirical analysis showed that kin discrimination is most likely to be favored where promiscuity occurs at intermediate levels. Another theoretical report confirmed that cooperative breeding can evolve from states displaying low levels of promiscuity combined with kin discrimination (Leggett et al. 2011), and Nielsen et al. (2012) demonstrated significant deleterious effects from inbreeding in cooperatively breeding meerkats (*Suricata suricatta*), a condition that would constrain social evolution without kin discrimination.

Assuming that the previous patterns generalize to mammals, they suggest two paths to cooperative breeding, one "dependent on the presence of helpers" (callimicos?: Fig. 5.1), another in which cooperative breeding is "facultative, with some pairs breeding successfully without helpers" (maras?: Fig. 5.1). Because mammalian evolution was strongly influenced by heterogeneous (unpredictable, stressful) regimes, favoring flexible, opportunistic decision making (Jones 2009; this brief, Chaps. 6 and 9), it might be expected that, among cooperatively breeding members of the class, "facultative" structures would be most common. On the other hand, if "ecological constraints" (Emlen 1982) are the primary determinants of breeding in natal groups and reduced dispersal from natal groups, cooperatively breeding species may be obligately dependent upon the presence of helpers.

Fig. 5.1 Callimico (*left*, Primates: Callitrichidae, *Callimico goeldii*) at Estacion Biologica Tahua-manu, Department of Pando, Bolivia, eating jelly fungus, *Auricularia delicate*. After Porter and Garber (2009), Callimicos are small (adults ~355–366 g), clawed, primarily arboreal monkeys that, unlike other, twinning, callitrichids, produce singletons and reproduce twice per year. Callimico spatial dispersion grades from monogamous pairs (in captivity) to small, cohesive, cooperative single- or multimale-multifemale units in which all females are reproductives. Callimico group structure may be intermediate between that of mara (*right,* Rodentia: Caviidae, *Dolichotis patagonum*), exhibiting monogamy grading into communal nesting and reluctant, occasional nursing by females of other females' young (Campos et al. 2001), and that of cooperatively breeding mammals (Solomon and French 1997). (Callimico ©Leila Porter, Edilio Nacimento Becerra; Mara ©Claudia M. Campos)

5.2   Primitively Eusocial Mole Rats

Following Sherman et al. (1991), Lacey et al. (2000), Lacey and Sherman (1997), it is acknowledged that, among mammals, some species of mole rats (Bathyergidae: naked mole rats, *Heterocephalus glaber*; Damaraland mole rats, *Fokomys damarensis*) represent the only known examples of eusociality in the class. Though some morphological differentiation has been observed for naked mole rats, these eusocial rodents remain at the "primitive" stage because, like cooperatively breeding mammals, types are functionally and potentially totipotent. Sociality in mole rats is presumed to have derived via a solitary state or from monogamy in conditions severely compromising reproduction devoid of benefits from group life (communal nest, cooperative burrowing: see Alexander et al. 1991). The previous condition suggests that the evolutionary landscape from "sexual segregation" to monogamy and, then, to eusociality is very steep.

Naked mole rats exhibit at least one conservative feature characteristic of mara (Fig. 5.1): A female of a mated pair is resistant to nursing other females' young, demonstrating that, despite a number of traits analogous to social insects, the evolution of mole rat eusociality has been constrained by intrinsic or extrinsic factors. On the other hand, the eusocial status of naked and Damaraland mole rats may be the "best of a bad job" or, simply, good enough for reproductive individuals in challenging regimes, and extreme conditions may require extreme solutions. Indeed, loss of eyesight, virtually all bodily hair, and other adaptations (slow ageing) indicate that naked mole rats have reduced energy allocation to the bare minimum for all but feeding and reproductive functions. Like many cooperatively breeding mammals and social insects, dominant *H. glaber* and *F. damarensis* females emit pheromones to repress reproduction in subordinate females.

Crespi (1994) advanced three sufficient, but not necessary conditions for the evolution of eusociality : "(1) food-shelter coincidence, (2) strong selection for defense, and (3) ability to defend." The previous author emphasized ecological factors as determining factors for the evolution of eusociality, including, "extremely high value of the habitat," enhancing "possibilities for habitat inheritance, high relatedness in claustral situations, self-sufficiency of juveniles, greater ability of workers to reproduce, and trade-offs between defensive ability and dispersal." In addition to habitat factors, climate is implicated as a significant factor for bathyergid eusociality since naked and Damaraland mole rats are found in stressful, arid regimes with compacted soils, increasing burrowing costs and, possibly, increasing benefits of cooperation and/or altruism. Social mole rats match Crespi's (1994) three conditions, living virtually all of their lives underground eating an abundant, evenly available food (tubers), protected from predators via their extensive, multichambered burrow. The same author suggested that other eusocial mammals might be identified among burrowing taxa (hystricognath rodents?, gerbilline rodents?, moles?). Investigations are required to continue the search for additional eusocial mammals, to define evolutionary routes to eusocial societies in the class, and to compare and contrast the various grades of mammalian sociality, including the number of times it has arisen independently in the class.

5.3   Toward a Social "Toolkit"

Components of a social "toolkit" mapping pathway feedbacks, from gene to phenotype to environment and back, require theoretical and empirical tests in the context of the evolution of mammalian population structures, from solitary to eusocial. In particular, adopting a comparative, functional trait-based approach (Fig. 2.3; McGill et al. 2006) to the study of mammalian social evolution will enhance projects seeking generalities across the class. Additional components of the "toolkit" may be particular morphological and behavioral features such as "driving" (Eisenberg 1981) and "retreat-expansion" (Tanaka 1973), formations adopted during dyadic interactions. Among mammals, "driving" postures are ubiquitously observed during precopulatory and courting phases of mating, and "retreat-expansion" or "concentration-dispersion" (Lee 1976) formations are generic as approach–avoid, fight–flight, or bold–shy action patterns.

Among extant mammals, the idea of a mammalian "toolkit" is promoted by the finding that "within-group relatedness and allomaternal care are positively correlated and conserved throughout the mammalian phylogeny" and that female mammals, in general, demonstrate tolerance for non-relatives (Briga et al. 2012). A social "toolkit" may represent a quantifiable "map" of genetically correlated traits and of diversity available to selection (Table 3.1), highlighting one advantage of having a high reproductive rate to expose a type to the environment via the phenotypes of descendants. Environmental heterogeneity and extreme stress were significant evolutionary effects in mammals (*Morganucodon, Kuehneotherium:* Kermack and Kermack 1984; Soricidae: Churchfield 1990; Hawes 1977; this brief, Chap. 3), the flexibility of their phenotypes is well documented and highly touted, and many taxa are genetically monomorphic (Selander and Kaufman 1973; Jones 2012; but see Nielsen et al. 2012).

It is widely claimed that mammals are characterized by "generalized" phenotypes adapted to heterogeneous regimes (see Jones 2009; cf. Kermack and Kermack 1984), and those historical conditions may have favored the "toolkit" outlined above, and the literature on the evolution of specialist and generalist strategies provides a gateway to understanding mammalian social evolution. Discussing "the interplay between ecological and evolutionary dynamics," Cameron et al. (2013) and Pandit et al. (2009) discuss literature showing that variations in population density and population structure mediate life history via selection of genotype-environment interactions, favoring one or more "functional traits". Dense populations are often associated with generalized "functional traits" while low population densities are often associated with specialized tactics and strategies (Fortin et al. 2008; Pandit et al. 2009). Specialized traits such as burrowing or hibernation indicate "coarse-grained" conditions (environmental perturbations longer than generation time), while generalized traits (caching of food, reversible physiological and behavioral responses), are associated with "fine-grained" conditions (environmental perturbations shorter than generation time; see Fig. 2.2 in Jones 2012). Many small, nocturnal mammals conform to the former state, many large mammals, to the latter.

The report by Pandit et al. (2009) suggests a link between generalist strategies and mammalian "sociality", not only because of traits favored by life in dense populations but, related to the latter, generalists are most sensitive to spatial effects, consistent with behavioral ecological models of aggregation and "sociality" (Chap. 6). The analysis provided by Martin et al. (2013) showed that environmental heterogeneity, in the present case, spatial variability, "can shift trait expression and mask trade-offs by reversing fitness consequences within species." Thus, interactions among the factors discussed in this paragraph may be sufficient to favor aggregation, "sociality," and varieties of facilitation (Fig. 2.1), suggesting that the drivers of the spatiotemporal effects discussed by Emlen and Oring (1977) may be more complex than they appear on surface.

Importantly, Chaianunporn and Hovestadt (2012) found that "mutualism tends to reduce dispersal in both partners," further evidence that tolerant strategies have the potential to reduce reproductive costs (Bonte et al. 2012), and, probably, the potential to facilitate the evolution of sociality by favoring kin groups. The aforementioned speculations require theoretical treatment, in particular, testing possible associations between the evolution of plastic phenotypes and the evolution of sociality, on surface, an apparently tenuous relationship (Table 3.1). A final *caveat* to the unqualified acceptance of spatiotemporal models of "social" evolution is exemplified by experiments with small herbivores (Stahl et al. 2006), including hares (*Oryctolagus cuniculus*), showing that competition may coexist with facilitation, conditions likely applicable to many mammals with highly developed nervous systems designed, in part, for opportunistic, more-or-less accurate discrimination and decision making (e.g., carnivores, primates, cetaceans).

5.4 Mammalian Sociality and Social Insects: Convergent Patterns Emerge

My review of primitively eusocial mammals demonstrates several patterns differentiating them from social insects and cooperatively-breeding birds. Evolution to advanced sociality appears to be constrained in primitively eusocial mammals because of shifts in reproductive optima occasioned by heterogeneous environments, with associated costs of competition for limiting resources. The latter conditions will increase conflicts of interest among group members and will increase benefits of repressive tactics ("policing," aggression, exploitation). Thresholds of r are difficult to measure. However, though coefficients of relatedness may be sufficiently high to favor cooperative-breeding where taxa inhabit heterogeneous regimes, the large number of "solitary" mammalian species suggests that environmental heterogeneity has more often favored dispersal from natal groups in the class, particularly, emigration of males. Thus, adaptation to variable conditions retains phenotypic flexibility among many large mammals, with the trade-off being more efficient and robust division-of-labor. For most large mammals, robustness is achieved via body size and, for many species, a high tolerance for genetic homogeneity (see Jones 2012).

Large brain-to-body size ratios (Armstrong 1983), as well, buffer mammals from environmental stochasticity, permitting types to track predictable aspects of local regimes. Brains, as well, provide benefits from opportunistic decision making and problem solving in regard to resource defense or resource access convertible to direct or indirect offspring. An expanded frontal cortex, also, incurs some advantages of both specialized and generalized tactics and strategies via accommodation to variable conditions (opportunistic "decision" making). The foregoing patterns suggest that social evolution is characterized by lineage-specific as well as general factors (Sect. 8.1, Synopsis), supporting the perspective that systematic research within and between classes will prove productive though, significantly, Fischman et al. (2011), studying "whole genome" comparisons of social insects employing "protein-coding sequence" analyses, identified "five major biological processes" (chemical signaling, brain development and function, immunity, reproduction, as well as, metabolism and nutrition) positioned for comparative studies across social insect and vertebrate eusocial taxa.

References

Alexander RD, Noonan KM, Crespi BJ (1991) The evolution of eusociality. In: Sherman PW, Jarvis JUM, Alexander RD (eds) The biology of the naked mole-rat. Princeton University Press, Princeton

Armstrong E (1983) Relative brain size and metabolism in mammals. Science 220:1302–1304

Bonte D, Van Dyck H, Bullock JM, Coulon A, Delgado M, Gibbs M, Lehouck V, Matthysen E, Mustin K et al (2012) Costs of dispersal. Biol Rev 87:290–312

Bourke AFG (2011) Principles of social evolution. Oxford University Press, Oxford

Briga M, Pen I, Wright J (2012) Care for kin: within-group relatedness and allomaternal care are positively correlated and conserved throughout the mammalian phylogeny. Biol Lett 8:533–536

Cameron TC, O'Sullivan D, Reynolds A, Piertney SB, Benton TG (2013) Eco-evolutionary dynamics in response to selection on life-history. Ecol Lett 16:754–763

Campos CM, Tognelli MF, Ojeda RA (2001) *Dolichotis patagonum*. Mamm Spec 652:1–5

Chaianunporn T, Hovestadt T (2012) Evolution of dispersal in metacommunities of interacting species. J Evol Biol 25:2511–2525

Churchfield S (1990) The natural history of shrews. Cornell University Press, Ithaca

Cornwallis CK, West SA, Griffin AS (2009) Routes to indirect fitness in cooperatively breeding vertebrates: kin discrimination and limited dispersal. J Evol Biol 22:2445–2457

Cornwallis CK, West SA, Davis KE, Griffin AS (2010) Promiscuity and the evolutionary transition to complex societies. Nature 466:969–972

Crespi BJ (1994) Three conditions for the evolution of eusociality: are they sufficient? Insectes Sociaux 41:395–400

Crespi BJ, Choe JC (1997) Explanation and evolution of social systems. In: Choe JC, Crespi BJ (eds) The evolution of social behavior in insects and arachnids. Cambridge University Press, London

De Jong G (1976) Selection always increases efficiency. Am Nat 110:1013–1027

Eisenberg, J.F. (1981). The mammalian radiations: an analysis of trends in evolution, adaptation, and behavior. University of Chicago Press, Chicago

Emlen ST (1982) The evolution of helping. I. An ecological constraints model. Am Nat 119:29–39

Emlen ST, Oring LW (1977) Ecology, sexual selection, and the evolution of mating systems. Science 197:215–223

Fischman BJ, Woodard SH, Robinson GE (2011) Molecular evolutionary analyses of insect societies. Proc Nat Acad Sci USA 108(Supplement 2):10847–10854

Fortin D, Morris DW, McLoughlin PD (2008) Habitat selection and the evolution of specialists in heterogeneous environments. Il J Ecol Evol 54:311–328

Hawes ML (1977) Home range, territoriality, and ecological separation in sympatric shrews, *Sorex vagrans* and *Sorex obscurus*. J Mammal 58:354–367

Jones CB (2009) The effects of heterogeneous regimes on reproductive skew in eutherian mammals. In: Hager R, Jones CB (eds) Reproductive skew in vertebrates: proximate and ultimate causes. Cambridge University Press, Cambridge

Jones CB (2012) Robustness, plasticity, and evolvability in mammals: a thermal niche approach. Springer, New York

Kermack DM, Kermack KA (1984) The evolution of mammalian characters. Croom Helm, London

Lacey EA, Sherman PW (1997) Cooperative breeding in naked mole rats: implications for vertebrate and invertebrate sociality. In: Solomon NG, French JA (eds) Cooperative breeding in mammals. Cambridge University Press, New York

Lacey EA, Patton JL, Cameron GN (eds) (2000) Life underground: the biology of subterranean rodents. The University of Chicago Press, Chicago

Lee RB (1976) !Kung spatial organization: an ecological and historical perspective. In: Lee RB, DeVore I (eds) Kalahari hunter-gatherers: studies of the !Kung San and their neighbors. Harvard University Press, Cambridge

Leggett HC, El Mouden C, Wild G, West S (2011) Promiscuity and the evolution of cooperative breeding. Proc Roy Soc Lond B 279:1405–1411

Martin BT, Jager T, Nisbet RM, Preuss TG, Grimm V (2013) Predicting population dynamics from the properties of individuals: a cross-level test of dynamic energy budget theory. Am Nat 181:506–519

McGill BJ, Enquist BJ, Weiher E, Westoby M (2006) Rebuilding community ecology from functional traits. Trends Ecol Evol 21:178–185

Nielsen JF, English S, Goodall-Copestake P, Wang J, Walling CA, Bateman AW, Flower TP, Sutcliffe RL et al (2012) Inbreeding and inbreeding depression of early life traits in a cooperative mammal. Mol Ecol 21:2788–2804

Pandit SN, Kolasa J, Cottenie K (2009) Contrasts between habitat generalists and specialists: an empirical extension to the basic metacommunity framework. Ecology 90:2253–2262

Porter LM, Garber PA (2009) Social behavior of callimicos: mating strategies and infant care. In: Ford SM, Porter LM, Davis LC (eds) The smallest anthropoids: the marmoset/callimico radiation. Springer, New York

Russell AF, Sharpe LL, Brotherton PNM, Clutton-Brock TH (2003) Cost minimization by helpers in cooperative vertebrates. Proc Nat Acad Sci USA 100:3333–3338

Saltzman W (2003) Reproductive competition among female common marmosets (*Callithrix jacchus*): proximate and ultimate causes. In: Jones CB (ed) Sexual selection and reproductive competition in primates: new perspectives and directions. American Society of Primatologists, Norman

Selander RK, Kaufman DW (1973) Genic variability and strategies of adaptation in animals. Proc Nat Acad Sci USA 70:1875–1877

Sherman PW, Jarvis JUM, Alexander RD (eds) (1991) The biology of the naked mole-rat. Princeton University Press, Princeton

Sherman PW, Lacey EA, Reeve HK, Keller L (1995) The eusociality continuum. Behav Ecol 6:102–108

Solomon NG, French JA (eds) (1997). Cooperative breeding in mammals. Cambridge University Press, New York

Stahl J, Van Der Graaf AJ, Drent RH, Bakker JP (2006) Subtle interplay of competition and facilitation among small herbivores in coastal grasslands. Funct Ecol 20:908–915

Tanaka J (1973) Social structure of the Bushmen. In: Carpenter DR (ed) Behavioral regulators of behavior in primates. Bucknell University Press, Lewisburg

Wilson EO (1971) The insect societies. Harvard University Press, Cambridge, MA

Woodard SH, Fischman BJ, Venkat A, Hudson ME, Varala K, Cameron SA, Clark AG, Robinson GE (2011) Genes involved in convergent evolution of eusociality in bees. Proc Nat Acad Sci USA 108:7472–7477

Chapter 6
Ecological Models as Working Paradigms for "Unpacking" Positive and Negative Interactions Among Social Mammals

No level of ecological benefit can bring about altruism if
relatedness is not above zero.

(Bourke 2011)

Aggregation should reduce the predation risk of foragers by
dilution of risk, group defense, or increased vigilance.

(Street et al. 2013)

Abstract Group formation and group maintenance are necessary precursors to the evolution of social behavior, and spatiotemporal models remain the primary explanations for the two previous processes. In brief, where limiting resources are clumped in time and space, occasioned in heterogeneous regimes, female dispersion (distribution and abundance) "maps" onto resource abundance, and reproductive male dispersion "maps" onto that of females. It has been suggested that sexual selection (intersexual and intrasexual selection) provides the "glue" holding groups together, resulting in stable groups. This chapter discusses those topics as well as the roles played by predation and interspecific competition in the formation and maintenance of groups. As demonstrated in Chap. 2, however, ecological factors alone are not sufficient to effect a transition from aggregations to societies since Hamilton's rule must attend.

Keywords Ecological models · Aggregation · Predation · Chesson's model · Competition · Tradeoffs · "Stress-gradient" hypothesis

Group formation and group maintenance are necessary precursors to the evolution of social behavior. Sociality may evolve where an organism's reproductive benefits from facilitating the reproduction of a conspecific, especially a relative, outweigh benefits without such facilitation. Relying upon frameworks developed by behavioral ecologists, this section addresses group formation ("aggregations": Hamilton 1971; Street et al. 2013) and spatiotemporal maintenance of groups (Emlen and Oring 1977; Bradbury and Vehrencamp 1977) as necessary, but not sufficient or inevitable, preconditions for social evolution. Across taxa, mainstream literature supports the idea that transitions to social states, should they occur, are deterministic functions of limiting resource dispersion (Emlen and Oring 1977; Emlen 1982), predation

C. B. Jones, *The Evolution of Mammalian Sociality in an Ecological Perspective*,
SpringerBriefs in Ecology, DOI 10.1007/978-3-319-03931-2_6,
© Clara B. Jones 2014

(Crespi 1994; Strassmann et al. 1988; Ebensperger and Blumstein 2006), inter-specific competition (Orians and Willson 1964; Moynihan 1968; Schoener 1982), or some combination of these and other variables (e.g., repression of aggression, repression of competition: Wilson 1971, 1975; Brown 1975; Wheeler 1928). However, for each hypothesized mechanism, condition-dependent strategies, as well as differential reproductive costs, and benefits, have not been diagnosed, particularly, in relation to variations in age (reproductive value), sex, and reproductive condition (but, see, Lehmann and Keller 2006). Vehrencamp (1979: "routes," this brief, Chap. 1), Helms Cahan et al. (2002: "trajectories"), and Bourke (2011: subsocial and semisocial "pathways"), and Crespi (2007: sociosexual coevolution, this brief, Chap. 1) advanced verbal models incorporating the concept of evolutionary stages. However, the hypothesized preconditions have rarely been tested experimentally.

6.1 The Behavioral Ecology of Group Formation and Stable Maintenance of Groups

From microbes to mammals, ecological factors correlate with group formation (e.g., "forage-mediated aggregation": microbes: Kadam and Velicer 2006; Lazazzera 2000; Crespi 2001, insects: Crespi 1994; Wilson and Hölldobler 2005; Trumbo 2009; Strassmann 1981, crustaceans: Toth and Duffy 2004, amphibians: Arak 1983, fish: Hoare et al. 2004, birds: Fleming et al. 1987; Davies 1992, mammals: Solomon and Getz 1997, p. 223, Eisenberg 1966, 1981; Clutton-Brock and Harvey 1978; Wrangham 1979; Fisher and Owens 2000; Benoit-Bird and Würsig 2004, population differentiation: Hoelzel et al. 1998, 2007; Xing et al. 2009, speciation: Rundle and Nosil 2005; Jones 2013; see Packer et al. 1990). To date, the most powerful and widely accepted models of social evolution propose that adaptations for group living are favored in stochastic environments characterized by spatiotemporally "patchy," clumped, limiting resources (e.g., food, mates, breeding sites), onto which the dispersion of mates is "mapped" (Crook 1964, 1965, 1970; Brown 1975; Wilson 1971, 1975; Emlen and Oring 1977, emballonurids: Bradbury and Vehrencamp 1977; Eisenberg 1966, 1981, cooperatively breeding *Suricata suricatta*: Bateman et al. 2013, ungulates: Street et al. 2013). Studying gregarious gerbils (*Gerbillus allenbyi, G. pyramidum*), Fierer and Kotler (2000) provided evidence that mammals are able to discriminate "patch" boundaries and to subdivide patches into "micropatches." Using an experimental design in seminatural conditions, Rémy et al. (2013) demonstrated differential effects of food dispersion on female bank voles (*Myodes glareolus*) when patterns of resource distribution and predictability were manipulated. The latter authors concluded, "These results suggest a tight relationship between the spatiotemporal distribution of food, social organization, and population dynamics." What are the details of this spatiotemporal model, and what accounts for its predictive power?

   In brief, first principles of ecology (energy acquisition, consumption, and allocation) support the view that the size and composition of groups change in response to spatiotemporal environmental heterogeneity (e.g., resource distribution, temperature, rainfall: Fig. 1.1), with attendant consequences for the survival and fecundity of organisms (e.g., Pulliam and Caraco 1984; Wang et al. 2006; Jones 2012). In 1982, Emlen revised the 1977 model formulated with Oring, emphasizing a need to explain the occurrence of cooperatively breeding reproductive units across a wide range of environmental conditions (units in which some group members forego reproduction to help one or more female rear offspring: Chap. 5). His "ecological constraints" model proposed that intense competition for limiting resources, particularly, breeding sites, may decrease benefits or increase costs of dispersal from natal groups, yielding high within-group relatedness, a propitious condition for social evolution via a subsocial pathway. Emlen (1982) explained how certain social structures evolve *after* stable groups have formed; thus, the 1982 treatment is not an alternative to Emlen and Oring (1977), a verbal model providing explanations for the origins of aggregations and stable groups.

Ecological models hold that population dispersion and structure are attributes of resource predictability (e.g., Emlen and Oring 1977; Bradbury and Vehrencamp 1977; Wilson 1975; Pulliam and Caraco 1984; Wong 2011). High resource predictability and quality, relatively homogeneous spatial dispersion of limiting resources, combined with resource tracking by the animal population, are expected to favor resource defense (e.g., contest competition or territoriality) by individuals or small groups (op. cit.; Davies et al. 2012). However, low resource predictability and large distance or high variation in distance between resource patches may make resources indefensible (not monopolizeable: DeYoung et al. 2009), yielding large average group size (Emlen and Oring 1977; Bradbury and Vehrencamp 1977; Pulliam and Caraco 1984; Crook 1964, 1965; Schoener 1971). Since temporal unpredictability of resources may be positively correlated with spatial uncertainty ("patchiness"), foraging in groups may reduce average search time per individual group member (see Handegaard et al. 2012). Thus, environmental predictability will be inversely correlated with group size (Wittenberger 1980; Pulliam and Caraco 1984; Wilson 1975), presumably, up to an asymptote determined by costs associated with high rates of interaction. In ecological models of social evolution, parameters determining modal group size in a population are ultimately expressed as adaptations of individuals to local conditions (Pulliam and Caraco 1984; Wilson 1975; Brown 1975; Wittenberger 1980; Hamilton 2010, this brief, Chaps. 3–5). Caveats obtain to certain fundamentals of spatiotemporal models because, for some mammalian taxa, food type and quality are significant determinants of population structure (e.g., Bodmer 1990), operating in association with resource dispersion, and, possibly, constraining the evolution of group formation or group maintenance.

6.2   Predation May Facilitate Group Formation, a Necessary Precursor to Social Evolution

Predation and interspecific competition are community-level effects (Chesson 2000; Chesson and Kuang 2008; Schoener 1974, 1982; Cody 1974; Garber 1988; Valone and Brown 1995; Ebensperger and Blumstein 2006, this brief, Chap. 2) whose roles in the proposed transitions from formation of aggregations to social evolution are poorly understood. In 1971, Hamilton proposed a "nearest-neighbor strategy" whereby aggregations form in response to predation pressure whose effective risk to a single individual would be diluted by joining a group (the "selfish herd"). The "dilution effect" has received empirical support in field and laboratory studies (Foster and Treherne 1981; James et al. 2004; Street et al. 2013). However, spatial scale and differential tradeoffs between foraging strategies and predator defense have been addressed theoretically, suggesting that selection intensities generated by predation may not be sufficiently reliable or directional to maintain aggregations once they have formed. Theoretical research has shown, as well, that the benefits of grouping in response to predation pressures are often unpredictable or seasonal rather than recurrent (Kie 1999; Nonacs and Blumstein 2010).

Conducting research on Serengeti cheetahs (*Acinonyx jubatus*), Durant et al. (2004) documented factors responsible for group formation among juveniles, including patterns of juvenile mortality associated with predation by lions (*Panthera leo*). Like most other felids (and most mammals), cheetah population structure is generally considered "solitary," exhibiting segregation (spatial differentiation) among adult males and females outside the breeding season, as well as relatively limited postnatal investment by females in their young. Durant and her colleagues documented the protective effects of "subsocial" architecture among siblings from the same litter. In particular, "adolescent male survival was strongly related to the presence of a sister but was unrelated to the presence of a brother." Differential effects of mortality were reported for adult males, also, since survivorship, and, reproductive success, were significantly higher for brothers forming lifelong coalitions. Interestingly, Durant et al. (2004) reported that group size was not associated with juvenile or male survival or mortality, suggesting that effects of selection pressures imposed by predation are independent of group size in these conditions.

6.3 Interspecific Competition May Facilitate Mammalian Group Formation, a Necessary Precursor to Social Evolution

Interspecific competition occurs when individuals of different taxa compete for the same limiting resource (food, mates, space). Generally, these interactions do not lead to "elimination" ("competitive coexistence": Chesson 2000; Schoener 1974, this brief, Chap. 2) due to evolved tactics and strategies minimizing the degree of spatial or temporal overlap between and among competitors, particularly, "resource partitioning" (Chesson 2000, this brief, Chap. 2). Interspecific competition may be effective above a critical threshold of population density and/or functional similarity of species, and, in these conditions, inclusive fitness outcomes may be positive (facilitation, mutualism), negative (competitive exclusion, local extinction), or neutral (tolerance), with outcomes determined by species exhibiting greater "resource-holding potential" (superior or dominant taxa) in a given "local" regime or thermal niche. The more closely related are individuals within and between populations, the more intense competition may be in some regimes, due to more similar demands for resources, including space and time.

Studying partitioning of resources by prey size among seven species of mustelid carnivores in Ireland, (weasel, *Mustela nivalis*: solitary, seasonally nocturnal; stoat, *M. erminea*: primitively polygynous, mostly nocturnal; mink, *Mustela lutreola*: solitary, mostly nocturnal; pole cat, *M. putorius*: solitary, nocturnal; pine marten, *Martes martes*: solitary, nocturnal; badger, *Meles meles*: solitary or clans, fossorial, nocturnal; otter, *Lutra lutra*: dens, primitively polygynous, territories strongly correlated with food availability), McDonald (2002) reported effects *x* sex (males' prey were larger than females') and *x* body size (larger predators ate larger prey) but not x prey size (no relationship between predator size and prey size). Though vermivorous badgers and piscivorous otters were sometimes outliers, the seven taxa were characterized by broad niches. Niches of these seven mustelids in Ireland were more similar than niches of the same species in Great Britain, highlighting the importance of ecological constraints in assembling communities as well as populations, the focus of Emlen and Oring's (1977) model. Because McDonald (2002) did not find that these mustelids partitioned prey by size, he concluded that interspecific aggression rather than resource partitioning may explain the observed patterns of assembly. These subjects deserve further investigation since the latter condition does not predict group maintenance, though the latter author's findings may be consistent with Chesson's (2000) formulations (i.e., intraspecific competition may be more intense in this study than is interspecific competition). This group might be a good model for the study of constraints on social evolution and transitions to group formation from a solitary, nocturnal state to male territories that overlap with those of females (stoats, otters), a primitive type of polygyny. Some of these species demonstrate a gradual tolerance for diurnal habits, a character worthy of investigation since diurnality and sociality are correlated in mammals.

6.4   How Robust are Predation and Interspecific Competition as Conditions for the Evolution of Group Maintenance, Possibly Leading to Social Evolution in Mammals

Relationships between social phenomena at the population level and in other dynamic (community- and ecosystem-level) processes have received insufficient attention (Cramer and May 1972; Bengtsson 1989). Nonetheless, functional similarities between predation and interspecific competition, as well as, similar effects on population responses, have been discussed (e.g., "defensive mimicry": Malcolm 1990; Cody 1969). Diagnosing the relationship between interspecific competition and group maintenance may prove more challenging than the association between predation and group maintenance since the former process often leads to increased dispersion of conspecifics rather than their aggregation. For example, highlighting the importance of "indirect interactions" among more than two species, Porter and Garber

(2007) (also see Vencl 1977; Schreier et al. 2009) reported interspecific competition for food-induced variations in grouping patterns among callimicos (*Callimico goeldii*) and two tamarin species (*Saguinus fuscicollis, S. labiatus*), all members of the neotropical primate family, Callitrichidae. When foraging in "polyspecific associations," callimicos exhibited dietary shifts requiring larger daily ranges over areas including more habitat types compared to patterns of resource use when feeding independent of competition from tamarins. The results of Porter and Garber (2007) demonstrate ways in which "indirect interactions" may obscure direct causes and effects of grouping patterns induced by interspecific interactions and their potential for social evolution (cf. Schoener 1974). The study by Porter and Garber (2007) suggests that certain assumptions of the spatiotemporal model are more complex than they appear on surface—phenomena in need of systematic investigation.

Robotics has been used to address questions about interspecific competition in taxa other than mammals, but with assumptions and methods having general applicability. Following the previous discussion, it was shown that interspecific competition may decrease rather than increase group cohesiveness ("attraction" or "polarization": Ioannou et al. 2012; Tinbergen et al. 1967), and predation may have the same effects. Using high-resolution ("acoustic video") imaging sonar, Handegaard et al. (2012) studied "coordinated group hunting" by spotted sea trout (*Cynoscion nebulosus*) preying on juvenile Gulf menhaden (*Brevoortia patronus*), research yielding insights into the destabilizing effects of predation. Handegaard et al. (2012) showed that certain features of predator groups (directionality, degree of coordination) corresponded with "incoherence" or "impedence" of schooling prey. By analogy, where interspecific competition operates, a dominant species may disrupt or interfere with mechanisms maintaining group coherence and coordination in a second species. These conditions were shown to increase per capita risk for menhaden prey in small schools. Predation and interspecific competition, then, may increase group dispersion, suggesting that either process may interfere with mechanisms of group cohesion (e.g., auditory, nonauditory, or olfactory communication: see Moynihan 1968; Cody 1969). Indeed, Handegaard et al. (2012) demonstrated that trout disrupted communication among prey by increasing their interindividual distance and decreasing behavioral coordination among members of menhaden schools. These and related methods deserve increased consideration by mammalogists for the investigation of group formation, having the potential, not only, to address seemingly intractable problems, but, also, to afford relatively noninvasive quasi-experimental and experimental designs.

Researchers testing propositions discussed in this section, however, must keep in mind that spatiotemporal and related hypotheses for the evolution of "sociality" are not based on mammalian models but rather on the bird and amphibian literature. To date, literature in animal behavior has not been sufficiently evaluated for possible differences among Classes, a concern highlighted by the report by Fisher and Owens (2000) showing that macropods do not conform well to a spatiotemporal ecology model. Table 3.2 makes a similar point exemplified by cavies (Caviidae), reinforcing caveats raised throughout this section: behavioral ecology models of group formation and group maintenance may derive their predictive power from complex interactions of both correlational and causal factors.

References

Arak A (1983) Male-male competition and mate choice in anuran amphibians. In: Bateson PPG (ed) Mate choice. Cambridge University Press, Cambridge

Bateman AW, Ozgul A, Nielsen JF, Coulson T, Clutton-Brock TH (2013) Social structure mediates environmental effects on group size in an obligate cooperative breeder, *Suricata suricatta*. Ecology 94:587–597

Bengtsson J (1989) Interspecific competition increases local extinction rate in a neotropical system. Nature 340:713–715

Benoit-Bird KJ, Würsig B, McFadden CJ (2004) Dusky dolphin (*Lagenorhynchus obscurus*) foraging in two different habitats: active acoustic detection of dolphins and their prey. Mar Mamm Sci 20:215–231

Bodmer RE (1990) Fruit patch size and frugivory in the lowland tapir (*Tapirus terrestris*). J Zool 222:121–128

Bourke AFG (2011) Principles of social evolution. Oxford University Press, Oxford

Bradbury JW, Vehrencamp SL (1977) Social organization and foraging in emballonurid bats III: mating systems. Behav Ecol Sociobiol 2:1–17

Brown JL (1975) The evolution of behavior. W.W. Norton, New York

Chesson P (2000) Mechanisms of maintenance of species diversity. Ann Rev Ecol Syst 31:343–366

Chesson P, Kuang JJ (2008) The interaction between predation and competition. Nature 456:235–238

Clutton-Brock TH, Harvey PH (1978) Mammals, resources, and reproductive strategies. Nature 273:191–195

Cody ML (1969) Convergent characteristics in sympatric populations: a possible relation to interspecific territoriality. Condor 71:222–239

Cody ML (1974) Competition and the structure of bird communities. Princeton University Press, Princeton

Cramer NF, May RM (1972) Interspecific competition, predation, and species diversity: a comment. J Theor Biol 34:289–293

Crespi BJ (1994) Three conditions for the evolution of eusociality: are they sufficient? Insect Soc 41:395–400

Crespi BJ (2001) The evolution of social behavior in microorganisms. Trends Ecol Evol 4:178–183

Crespi BJ (2007) Comparative evolutionary ecology of social and sexual systems: waterbreathing insects come of age. In: Duffy JE, Thiel M (eds) Evolutionary ecology of social and sexual systems: crustaceans as model organisms. Oxford University Press, Oxford

Crook JH (1964) The evolution of social organization and visual communication in the weaver birds (Ploceinae), Behaviour Supplement X. Brill, Leiden

Crook JH (1965) The adaptive significance of avian social organization. Symp Zool Soc Lond 14:181–218

Crook JH (ed) (1970) Social behaviour in birds and mammals. Academic, London

Davies NB (1992) Dunnock behaviour and social evolution. Oxford University Press, Oxford

Davies NB, Krebs JR, West SA (2012) An introduction to behavioural ecology, 4th edn. Wiley-Blackwell, Oxford

DeYoung RW, Demarais S, Gee KL, Honeycutt RL, Hellickson MW, Gonzales RA (2009) Molecular evaluation of the white-tailed deer (*Odocoileus virginianus*) mating system. J Mammal 90:946–953

Durant SM, Kelly M, Caro T (2004) Factors affecting life and death in Serengeti cheetahs: environment, age, and sociality. Behav Ecol 15:11–22

Ebensperger LA, Blumstein DT (2006) Sociality in New World hystricognath rodents is linked to predators and burrow digging. Behav Ecol 17:410–418

Ebensperger LA, Rivera DS, Hayes LD (2012) Direct fitness of group living mammals varies with breeding strategy, climate and fitness estimates. J Anim Ecol 81:1013–1023

Eisenberg JF (1966) The social organizations of mammals. Handbüch der Zoologie 8:1–92

Eisenberg JF (1981) The mammalian radiations: an analysis of trends in evolution, adaptation, and behavior. University of Chicago Press, Chicago

Emlen ST (1982) The evolution of helping. I. an ecological constraints model. Am Nat 119:29–39

Emlen ST, Oring LW (1977) Ecology, sexual selection, and the evolution of mating systems. Science 197:215–223

Fierer N, Kotler BP (2000) Evidence for micropatch partitioning and effects of boundaries on patch use in two species of gerbils. Fun Ecol 14:176–182

Fisher DO, Owens IPF (2000) Female home range size and the evolution of social organization in macropod marsupials. J Anim Ecol 69:1083–1098

Fleming TH, Breitwisch R, Whitesides GH (1987) Patterns of tropical vertebrate frugivore diversity. Ann Rev Ecol Syst 18:91–109

Foster WA, Treherne JE (1981) Evidence for the dilution effect in the selfish herd from fish predation on a marine insect. Nature 293:466–467

Garber PA (1988) Diet, foraging patterns, and resource defense in a mixed species troop of *Saguinus mystax* and *S. fuscicollis* in Amazonian Peru. Behaviour 105:18–34

Gittleman JL, Thompson SD (1988) Energy allocation in mammalian reproduction. Am Zool 28:863–875

Hamilton IM (2010) Foraging theory. In: Westneat DE, Fox CW (eds) Evolutionary behavioral ecology. Oxford University Press, New York

Hamilton WD (1971) Geometry for the selfish herd. J Theor Biol 31:295–311

Handegaard NO, Boswell KM, Ioannou CC, Leblanc SP, Tjøstheim DB, Couzin ID (2012) The dynamics of coordinated group hunting and collective information transfer among schooling predators. Curr Biol 22:1213–1217

Helms Cahan S, Blumstein DT, Sundström L, Liebig J, Griffin A (2002) Social trajectories and the evolution of social behavior. Oikos 96:206–216

Hoare DJ, Couzin ID, Godin JGJ, Krause J (2004) Context-dependent group size choice in fish. Anim Behav 67:155–174

Hoelzel AR, Dahlheim M S, Sterm J (1998) Low genetic variation among killer whales (*Orcinus orca*) in the Eastern North Pacific and genetic differentiation between foraging specialists. J Hered 89:121–128

Hoelzel AR, Hey J, Dahlheim MS, Nicholson C, Burkanov V, Black N (2007) Evolution of population structure in a highly social top predator, the killer whale. Mol Biol Evol 24:1407–1415

Ioannou CC, Guttal V, Couzin ID (2012) Predatory fish select for coordinated collective motion in virtual prey. Science 337:1212–1215

James R, Bennett PG, Krause J (2004) Geometry for mutualistic and selfish herds: the limited domain of danger. J Theor Biol 228:107–113

Jones CB (2012) Robustness, plasticity, and evolvability in mammals: a thermal niche approach. Springer, New York

Jones CB (2013) Constraints on speciation in human populations: phenotypic diversity matters. Hum Biol Rev 2:263–279

Kadam SV, Velicer GJ (2006) Variable patterns of density-dependent survival in social bacteria. Behav Ecol 17:833–838

Kie JG (1999) Spatial foraging and risk of predation: effects on behavior and social structure in ungulates. J Mammal 80:1114–1129

Lazazzera BA (2000) Quorum sensing and starvation: signals for entry into stationary phase. Curr Opin Microbiol 3:177–182

Lehmann L, Keller L (2006) The evolution of cooperation and altruism—a general framework and a classification of models. J Evol Biol 19:1365–1376

McDonald RA (2002) Resource partitioning among British and Irish mustelids. J Anim Ecol 71:185–200

McNab BK (1980) Food habits, energetics, and the population biology of mammals. Am Nat 116:106–124

Malcolm SB (1990) Mimicry: status of a classical evolutionary paradigm. Trends Ecol Evol 5:57–62

Moynihan ML (1968) Social mimicry: character convergence versus character displacement. Evolution Int J org Evolution 22:315–331

Nonacs P, Blumstein DT (2010) Predation risk and behavioral life history. In: Westneat DF, Fox CW (eds) Evolutionary behavioral ecology. Oxford University Press, New York

Orians GH, Willson MF (1964) Interspecific territories of birds. Ecology 45:736–745

Packer C, Scheel D, Pusey AE (1990) Why lions form groups: food is not enough. Am Nat 136:1–19

Porter LM, Garber PA (2007) Niche expansion of a cryptic primate, *Callimico goeldii,* while in mixed-species troops. Am J Primat 69:1340–1353

Porter LM, Garber PA (2009) Social behavior of callimicos: mating strategies and infant care. In: Ford SM, Porter LM, Davis LC (eds) The smallest anthropoids: the marmoset/callimico radiation. Springer, New York

Pulliam HR, Caraco T (1984) Living in groups: is there an optimal group size? In: Krebs JR, Davies NB (eds) Behavioural ecology: an evolutionary approach. Sinauer Associates, Sunderland

Queller DC (1992) A general model for kin selection. Evolution Int J org Evolution 46:376–380

Rémy A, Odden M, Richard M, Tyr Stene M, Le Gaillard J-F, Andreassen HP (2013) Food distribution influences social organization and population growth in a small rodent. Behav Ecol. doi:10.1093/beheco/art029

Rundle HD, Nosil P (2005) Ecological speciation. Ecol Lett 8:336–353

Schoener TW (1971) Theory of feeding strategies. Ann Rev Ecol Syst 2:369–404

Schoener TW (1974) Resource partitioning in ecological communities. Science 185:27–39

Schoener TW (1982) The controversy over interspecific competition: despite spirited criticism, competition continues to occupy a major domain in ecological thought. Am Sci 70:586–595

Schreier BM, Harcourt AH, Coppeto SA, Somi MF (2009) Interspecific competition and niche separation in primates: a global analysis. Biotropica 41:283–291

Solomon NG, Getz LL (1997) Examination of alternative hypotheses for cooperative breeding in rodents. In: Solomon NG, French JA (eds) Cooperative breeding in mammals. Cambridge University Press, New York

Strassmann JE (1981) Parasitoids, predators, and group size in the paper wasp, *Polistes exclamans.* Ecology 62:1225–1233

Strassmann JE, Queller DC, Hughes CR (1988) Predation and the evolution of sociality in the paper wasp *Polistes bellicosus.* Ecology 69:1497–1505

Street GM, Weckerly FW, Schwinning S (2013) Modeling forage mediated aggregation in a gregarious ruminant. Oikos 122:929–937

Tinbergen N, Impekoven M, Franck D (1967) An experiment on spacing out as a defence against predators. Behaviour 28:307–321

Toth E, Duffy JE (2004) Coordinated group response to nest intruders in social shrimp. Biol Lett 1:49–52

Trumbo ST (2009) Age-related reproductive performance in the parental burying beetle, *Nicrophorus orbicollis.* Behav Ecol 20:951–956

Valone RG, Brown JH (1995) Effects of competition, colonization, and extinction on rodent species diversity. Science 267:880–883

Vencl R (1977) A case of convergemce in vocal signals between marmosets and birds. Am Nat 111:777–782

Vehrencamp SL (1979) The roles of individual, kin, and group selection in the evolution of sociality. In: Marler P, Vandenbergh JG (eds) Handbook of behavioral neurobiology: social behavior and communication, vol 1. Plenum, New York

Wang G, Hobbs NT, Boone RB (2006) Spatial and temporal variability modify density-dependence in populations of large herbivores. Ecology 87:95–102

Westneat DE, Fox CW (eds) (2010) Evolutionary behavioral ecology. Oxford University Press, New York

Wheeler WM (1928) Insect societies: their origins and evolution. Harcourt Brace, New York

Wilson EO (1971) The insect societies. Belknap, Cambridge

Wilson EO (1975) Sociobiology: the new synthesis. Belknap, Cambridge

Wilson EO, Hölldobler B (2005) The rise of the ants: phylogenetic and ecological explanations. Proceedings of the National Academy of Sciences 102:7411–7414

Wittenberger JF (1980) Group size and polygamy in social mammals. Am Nat 115:197–222

Wong MYL (2011) Group size in animal societies: the potential role of social and ecological limitations in the group-living fish, *Paragobiodon xanthosomus*. Ethology 117:1–7

Wrangham RW (1979) On the evolution of ape social systems. Soc Sci Info 18:334–368

Xing J, Watkin WS, Witherspoon DJ, Zhang Y, Guthery SL, Mara R, Mowry BJ, Bulayeva K, Weiss RB, Jorde LB (2009) Fine-scaled human genetic structure revealed by SNP microarrays. Genome Res 19:815–825

Chapter 7
Mechanisms Underlying the Behavioral Ecology of Group Formation

Temperature dependence of trophic interactions is driven by asymmetry of species responses and foraging strategy.

Dell et al. (2013)

Individual metabolism thus provides a powerful and deep theoretical foundation on which to build an understanding of ecological processes and phenomena from the individual to populations, communities, and ecosystems.

Kearney and White (2012)

Population dynamics of social, group-living species can differ markedly from those of solitary species comprising relatively homogeneous populations. Social structure, per se, can have dynamical consequences.

Bateman et al. (2013)

Abstract This chapter discusses topics related to spatiotemporal models, including energetic effects and the evolution of group structure. It is proposed that research on the "stress-gradient" hypothesis may become a program representative of "integrative" social biology, as indicated by recent studies using California mice, *Peromyscus californicus*, as subjects. Positive and negative effects of social mammals are addressed for their roles in community assembly, emphasizing "bottom-up," "top-down," and indirect effects. A case study of competitive coexistence between two taxa in Costa Rican riparian habitat is presented, demonstrating that mammalian species may, at once, be inferior or mutualistic with members of other Classes, particularly, insects and birds, respectively. Topics discussed in this chapter pertain, as well, to conservation of species diversity.

Keywords Behavioral Ecology · Group formation · Aggregations · Group maintenance · Density dependence · Facilitation

This chapter discusses feedbacks between ecological factors and populations of social mammals, including effects at various scales of organization within ecosystems (Fig. 1.4). The type (genotype –phenotype) continues to be the basic unit of organization, making condition-dependent, self-interested "decisions" that may impact groups, populations, and communities. The role of social mammals in communities is emphasized, in particular, their participation in indirect interactions and their potential roles as "ecosystem engineers" as well as superior or inferior competitors. The present chapter is linked to spatiotemporal models of social evolution by the proposition that responses of populations vary along a continuum of stress (the "stress-gradient hypothesis") whereby groups are more likely to form in stressful conditions (conditions dampening reproductive rates), presumably because grouping, initially, conserves energy. A variety of roles played by social mammals in ecosystems is stressed, and their mechanisms and functions highlighted to show the benefits to types and their environments from groups once they have formed and become spatiotemporally recurrent. The topics in this chapter have direct import for the maintenance of biodiversity and the roles that social mammals play in the health of ecosystems (Fig. 1.4).

Applying Kearney and White's (2012) statement to the level of types, one might propose that metabolism drives interactions between types in overlapping thermal zones. Outcomes of interactions are determined by asymmetries in trait expression (frequency, rate, duration, intensity, quality) between types responding to environmental, including social, stimuli. Food is the fundamental driver of density dependence, and density dependence is a necessary factor for group formation. As density increases in a population, interaction rates of individuals often increase. Under these conditions, it is likely to benefit some individuals to exhibit tolerance of conspecifics (Table 2.1), perhaps as a by-product of competition for limiting resources, enhancing the reproductive interests of one or more types in the aggregation. In certain conditions, selection will favor types that are responsive to abiotic and biotic (including social) stressors (factors initially decreasing population growth rates; Sibley and Hone 2003) and to types with shared reproductive interests (Chap. 1). The "combined action of environmental stressors" defines a "thermal zone" within which each individual in a population behaves ("ecological space"). A type's ecological space defines a "set of points" within its population's thermal range, each point representing a thermal zone in which inclusive fitness is maximized (see Jennings and Veron 2011). It is in this sense that a type may contribute to (shifting) mean fitnesses of a population, and each type's phenotypes are expressed in interaction with its thermal zone, with relative measures of reproductive benefits and costs.

7.1 Behavioral Ecology, a Paradigm for the Evolution of Group Structure: Extrinsic Factors Shape "Decisions" Made by Types

Mechanisms and functions, causes and consequences of "facilitation" have been central to studies in plant and animal ecology, and ecological theories are central to studies of social evolution (e.g., group formation and group maintenance). Emlen's verbal models (Emlen and Oring 1977; Emlen 1982; this brief, Chap. 6), and extensions of it (Wrangham 1980; Sterck et al. 1997), are deterministic formulations emphasizing the potential for spatiotemporal variations in resource dispersion (heterogeneity, fluctuations) to induce group formation, group maintenance, and sociality. In studies of plant ecology and plant–herbivore associations, these abiotic and biotic effects have been formulated as the "stress-gradient hypothesis," positing two community-level processes along a continuum, facilitation and competition, suggested to covary with environmental stressors (extreme environmental effects on organisms) influencing levels of organization of organisms, from biochemical (e.g., gene expression) to whole-organism phenotype (e.g., behavior). Anticipating the development of general formulations in future, it seems important for students of mammalian social biology to incorporate measures of "desiccation" or "temperature" in their definitions of "stress", variables employed by plant ecologists.

Recent theoretical and empirical tests of "the facilitation–competition paradigm under the stress-gradient hypothesis" have demonstrated that these quantitative relationships may not be linear, straightforward, or expressed at all (Kawai and Tokeshi 2007; Daleo and Iribarne 2009; Holmgren and Scheffer 2010; le Roux and McGeoch 2010; Malkinson and Tielbörger 2010). For example, Daleo and Iribarne (2009; also see Kawai and Tokeshi 2007) showed that, under some conditions, facilitation between South American salt-tolerant grasses (*Spartina* spp.) and borrowing, herbivorous crabs (*Neohelice* spp.) were suppressed by variations in substrate quality. Importantly, Holmgren and Scheffer (2010) reported that, in some communities, "strong facilitation" is found in "mild environments," indicating that the expression of facilitation and competition are condition-dependent, thus, accompanied by life-history trade-offs (genotype × environment interactions). Research comparable to the aforementioned reports is needed for mammals.

Crook's (1964, 1965, 1970) "behavioral ecology" model, expanded and generalized by Emlen and Oring (1977; Bradbury and Vehrencamp 1977; Emlen 1982; see overview of likely precursors in Ricklefs 1977), is characterized in terms of "facilitation" and "competition" where "facilitation" includes variations in social population structure (Table 2.1). In such conditions, one or more types within or between groups facilitate the reproduction of a conspecific (Preface, Chaps. 1 and 2). Consistent with theoretical and empirical work in behavioral ecology, variations in social behavior and social organization are herein evaluated relative to variations in stress gradients and type × environment (including social environment) interactions. Paraphrasing Kawai and Tokeshi (2007), "the relative importance of [facilitation] and competition, conceptually formalized in the 'stress-gradient hypothesis', [predict] that the net negative competitive effects are more important under relatively benign environmental conditions, whereas positive facilitative effects are more important under harsher conditions." The latter view continues to represent a litany in behavioral ecology, though, as Bourke (2011) pointed out, no degree of environmental effect will direct the evolution of sociality unless Hamilton's rule is satisfied.

Testing theories of facilitation and competition under field conditions will be challenging for social biologists studying mammals. Theoretically, traits associated with facilitation should vary with heterogeneity of environmental stressors, switching from competition to facilitation along a stress gradient such as a temperature gradient from riparian to deciduous habitats in tropical dry forest. Jones (1985) found, for example, that "reproductive skew" (defined in this study as linearity of dominance hierarchies) for reproductive males within groups decreased from riparian (wetter) to deciduous (drier) habitat. Direct tests of the stress-gradient hypothesis are particularly important for studies of mammals because of the ubiquitous role of heterogeneous regimes historically and currently. Quantitative, including experimental, investigations of the expression and evolution of traits associated with reaction norms expressed in stochastic environments require measurement in order to assess predictions of the "stress-gradient hypothesis," for example, whether reaction norms reflect "negative environmental correlations" or "genetic trade-offs" (Jasmin and Zeyl 2012; Everett et al. 2012). *Caveats* are in order when seeking quantitative statements about environment × type interactions since Cody's (1974, pp. 160–161, 207) treatment suggests that tolerance and facilitation may be favored when conditions are at their best *and* their worst, an "inverted-U" function. Thus, the tolerance–facilitation curve may not be a direct relationship.

Research programs designed to test the stress-gradient hypothesis using mammals reveal certain ways in which traits influence variations in environmental variables (pack-living wolves, *Canis lupus* (Fortin et al. 2008); cooperatively breeding meerkats, *Suricata suricatta* (Bateman et al. 2013); cooperatively breeding tamarins, Callitrichidae (Bicca-Marques and Garber 2003); general theoretical article by Sterck et al. 2011). This domain of study has the potential to be truly integrative (Blumstein et al. 2010) as indicated by a report by Trainor et al. (2010) showing that variations in social aggression in monogamous male California mice (*Peromyscus californicus*) are associated with variations in kinase PERK (protein kinase RNA-like endoplasmic reticulum kinase), regulating neuronal activity (neural plasticity) and serving as a transcription factor. Kinase PERK activity is sensitive to environmental input. Thus, its responses are density-dependent, and, since the protein is a major component of negative feedback loops in generalized stress responses associated with at least 30 biological processes, its potential for multiple trade-offs is, most likely, significant. It appears that studies of the relationship between stress gradients and facilitation will become central to an understanding of mammalian social evolution.

7.2   Types Influence and Are Influenced by Abiotic and Biotic Regimes: Positive and Negative Effects Among Species

"Decisions" made by a social type have implications for its own reproductive rate, the reproductive rates of other types, and the shifting mean fitness of their population (Chap. 2). Basic models of population growth rate weight all individuals in a population equivalently (Sibley and Hone 2003). The rationale for the latter quantitative approach rests on the fact that evolution is a population-level phenomenon. Further, individual-level data are not available that would permit modeling of spatiotemporal changes in population density (Martin et al. 2013). The present brief's individual-level perspective on mammalian social evolution requires measurement of "reaction norms" as well as their contribution to and effects for shifting mean population fitness. The schema in Fig. 1.3 displays the options of "fitness-maximizing" individuals (Fig. 1.2) as functional components of biomes (Fig. 1.1), influencing and influenced by the ecological and evolutionary processes governing population, community, and ecosystem assembly (Figs. 1.1, 1.4). In particular, "top-down" (predators to nutrients) and "bottom-up" (nutrients to predators) effects (Moreau et al. 2006; Narwani and Mazumder 2013; Kerimoglu et al. 2013; Prevedello et al. 2013; this brief, Fig. 1.4) incorporate "decisions" (actions, strategies; Fig. 1.3) made by types governed by Hamilton's rule (Fig. 1.2), influencing their reproductive rates and, consequently, the Malthusian parameters of their populations (Lande 2007).

The importance of top-down effects has been amply demonstrated for group-living mammals, and, in some environmental regimes, predators ("natural enemies"; Chesson 2000) strongly influence coexistence by regulating differential values of R^* among species in a community (Chesson 2000; Chesson and Kuang 2008; this brief, Chap. 2). If predation is nonrandom by type, "decisions" are likely to be influenced *via* concomitant variations in l_{within} and l_{between}, determining population structure and likelihoods of group maintenance and formation. Studying pack-living dingos (*Canis lupus dingo*), the largest land predator in Australia, Letnic et al. (2011; Fleming et al. 2012) showed that the loss of this secondary consumer led to "an increase in the abundances and impacts of herbivores and invasive mesopredators" (monogamous red fox, *Vulpes vulpes*). As a consequence of the latter events, many small and medium-sized mammals disappeared and plant biomass suffered due to perturbations induced on herbivore populations, identifying dingos as a "biodiversity regulator" (Letnic et al. 2011). Teichman et al. (2013) provided documentation of a remarkable trophic cascade from Canadian ungulates to birds and butterflies *via* ungulates' effects on shrubs.

Social herbivores and carnivores may impact ecosystem functions, effects fundamentally driven by a type's "decisions" driven by Hamilton's rule and dynamic variations in l^*, states fundamental to group formation in original conditions. Marshall et al. (2013) identified complexity in top-down processes. The latter authors demonstrated that willow populations failed to recover after reintroduction of pack-living grey wolves (*Canis lupus*; Fortin et al. 2008) to Yellowstone National Park (USA). Willows had been decimated by elk browsing, and recovery of ecosystem stability proved to be dependent upon "restoring hydrological conditions" in addition to wolf reintroduction (Marshall et al. 2013; Prevedello et al. 2013). Multitrophic processes ("indirect interactions"; see Arthur and Prugh 2010) as exemplified previously, generally entail relations among, and consequences for, predators, herbivores, and primary producers. Herbivores may promote biodiversity (coexistence) by influencing plant growth, positive or negative effects that may, in turn, influence limiting resource dispersion, possibly, creating conditions favorable to group formation (see Sankaran et al. 2013). A key point is that predation may drive "decisions" to join or remain in groups and, beyond the apparent import of dispersion of prey, differential rates of predation may influence relative degrees of clumping and, thereby, benefits of facilitation to predator and prey (Table 2.1).

Top-down ("indirect") effects have also been demonstrated for monogamous coyotes (*Canis latrans*; Prugh 2005), and social mammals can be regulated as well as serve as regulators (polygynandrous lions, *Panthera leo* (Kissui and Packer 2004); colonial Dall's sheep, *Ovis dalli* (Arthur and Prugh 2010)). The aforementioned results are consistent with the conclusions of Schneider and Brose (2012) that "ecosystem functioning" is not a result of "simple effects of biodiversity" (coexistence mechanisms). The report by Rebollo et al. (2013) provides another perspective on the latter view, showing effects of small, noncolonial herbivores on the population structure and species diversity of large mammalian herbivores dominating grassland habitats. Further evidence that bottom-up and top-down effects are different than, and possibly interact with, other patterns of interspecific interactions was provided by Prugh and Brashares (2011) studying influences on community structure by the typically solitary but opportunistically "social" (i.e., aggregating) kangaroo rat (*Dipodomys ingens*), an ecosystem engineer. Further (theoretical) analysis of facilitation (e.g., sociality; Table 2.1) as a modulated as well as modulating force in the maintenance of biodiversity (species coexistence) was provided by Chesson and Kuang (2008) showing that "multitrophic" phenomena need to be considered, particularly synergistic effects between predation and competition. This perspective has implications for any treatment of the differential consequences of "stress" for individuals and populations, including the likelihood that a social type will be favored by selection.

In 1980, Slobodkin observed: "Margalef considers the total information present in an ecological community to equal the number of choices that must be made in specifying the complete condition of the community" (Slobodkin 1980). Social and nonsocial mammals are expected to exhibit differential probabilities of within- and between-species interaction rates (Chap. 2), and because of reproductive (allocation) costs associated with a social type (types comprising groups in populations), tolerant strategies may often be favored by selection (Fig. 2.1). Populations comprising some conditional threshold of social types may perform most efficiently where tolerance (between types or species (mutualism)) and, possibly, sociality (i.e., cooperation or altruism between types) characterize associations with selected conspecifics.

A brief case in point concerns potential effects of social herbivores on soil quality and, possibly, tree germination and growth (see De Jager and Pastor 2008; De Jager et al. 2009). Many mammals exhibit "latrine behavior," recurrently urinating and defecating at particular locations on their territories or home ranges. For example, Barja (2009) reported selective fecal marking by pack-living wild wolves (*Canis lupus*), demonstrating selectivity (nonrandom choice) by soil type and tree species. Similar patterns of behavior have been reported for other social mammals (some primates), and spatiotemporally recurrent urination and defecation by groups compared to solitaries may enrich soils, enhancing soil permeability and tissue growth above and below ground. "Latrine behavior" may have important consequences in stressful environmental conditions as suggested by Givnish's (1999) observation: "Soil fertility... is a complex function of temperature, rainfall, and substrate texture, chemistry, drainage, and oxygenation. Effectively infertile soils should favor heavily defended foliage and low tree diversity." Coordinated activities resulting from cumulative, self-interested "decisions" are in need of systematic investigation because of the potential for social mammals to serve important roles in direct (spread of N and other fertilizing nutrients) and indirect (increased abundance of resources for primary consumers, increased diversity and stability) interactions within communities and ecosystems.

Conducting a multiplot, exclusion experiment in the field, Murray et al. (2013) reported that the effects of ungulates' "nitrogenous wastes" vary with spatial scale and the "seasonal timing of ungulate impacts," concluding that these functional traits link "fine-scale and landscape-level ecological processes." This study should be replicated with other mammals with hypothesized impacts upon soils and foliage *via* "nitrogenous wastes." The aforementioned research results suggest, again, that group formation and group maintenance are not necessarily straightforward functions of resource dispersion and quality, though the report from Murray et al. (2013) may support the "stress-gradient hypothesis," highlighting variations in "spatial scale." The benefits of group life and, possibly, varieties of facilitation (Table 2.1) may depend upon multilevel causes and consequences, including effects below the soil surface. If the latter scenario is valid, then social biology must expand to incorporate a "multitrophic," multilevel approach (Fig. 2.1; Fig. 2.2) to the feedback loops influencing the growth rates of types driven by thermal requirements, by minimization of l^*, and by constraints imposed by Hamilton's rule.

Other responses by mammals are thought to facilitate plant and, possibly, insect biodiversity. For example, the behavioral ecology literature abounds with studies documenting the roles played by group-living mammals as seed dispersers (large "solitary" and group-living mammalian herbivores; Goheen et al. 2010), and it has been shown that herbivores may stimulate growth and reproduction in their plant food prey, having positive as well as negative effects on forage. In addition to the aforementioned work, a new domain of research presages fundamental knowledge illuminating the positive roles played by mammals in ecosystem processes. Takahashi and Takahashi (2013), studying Japanese black bears (*Ursus thibetanus japonicas*), demonstrated that movements of individuals or aggregations created light gaps in forests, facilitating seedling growth. This unfolding program will assess each component of whole-organism phenotypes as features with the potential to contribute in positive ways to one or more forest level. Other researchers have investigated the effects of disturbance by social mammals. For instance, Queenborough et al. (2012) showed that arborescent palms dampened damage to seedlings and saplings caused by foraging peccaries (Tayassuidae). Subcanopy interactions such as the latter, however, may negatively impact some plants yet positively impact others; outcomes require study as disturbances with the potential to influence community structure and ecosystem dynamics. Again, group-living mammals may influence (Table 2.1) and be influenced by the spatial distribution of other species as well as limiting resources, with differential effects upon and responses to R^*, $l^*_{between}$, and l^*_{within}.

Additionally, both small and large terrestrial mammals may impact soil aeration, leaf litter decomposition, seed and earthworm depth, or other processes below soil surface *via* movements through forests or along trails that may, in turn, influence diversity of woody species and, in tropical forests, lianas (Schnitzer and Carson 2001; Mielke 1977). Such effects may have import for interspecific interactions (coexistence) in a range of habitats where climate change desiccates soils (Marcy et al. 2013). Similarly, arboreal mammals may impact "canopy gaps" in a manner similar to that described by Takahashi and Takahashi (2013) for Japanese black bears. Additionally, the anatomical designs of any body part contacting abiotic and biotic components of forests should be investigated for their possible contribution to processes facilitating biodiversity or habitat quality, with necessary implications for fitness-maximizing tactics and strategies. For example, the design of claws, feet, and hooves may represent functional traits associated with effects of locomotion such as soil compaction and aeration as well as seedling depth. The demonstrated and hypothesized mutualistic associations between "solitary" or aggregated mammals and their abiotic and biotic regimes might be studied to reveal reproductive benefits to mammals gained from functional traits operating as bodily effluents, anatomical and morphological characters, and action patterns.

Studying responses of rodents and ruminants to climate impacts and ecosystem "simplicity," Stien et al. (2012) documented synchrony in and between these herbivores' population parameters that might determine certain coordinated events, such as foraging patterns. Another example from the literature on rodents entailed fluctuating cycles of population density of small "keystone herbivores" across Europe driven by climate. Cornulier et al. (2013) concluded that these changes may be general across small herbivore populations with potential to cause cascading effects on trophic webs across ecosystems since these taxa are "keystone prey for predators, alternative prey, and forage plants." Large-scale effects on ecosystems have been documented for "solitary" and group-living mammals in Africa (McNaughton et al. 1988), and it will be important to systematically investigate the precise details of how these taxa influence "energy flow and nutrient cycling" (Carnicer et al. 2012; Martin et al. 2013).

For example, Salgado-Luarte and Gianoli (2012) identified effects of herbivores on particular plant functional traits (photosynthetic characteristics, leaf area, chlorophyll properties). Related to the aforementioned study, Holzwarth et al. (2012) documented "six different individual tree mortality modes" that require intense study relative to the potential of mammal groups to nonrandomly influence plant population parameters, including the interaction between "diet selection" and "the quantity and quality of food items" (Agrawal and Klein 2000). Determining how individual mammals assess the previous conditions and their reproductive costs and benefits ("reaction norms") may reveal similar patterns of decision-making, reflected in responses (Fig. 2.1), within and between "ecological niches" (Fig. 1.1), including no doubt the ubiquitous, multitrophic feedback loops impacting types and their fitness-maximizing decisions. Collection of data, based on field and laboratory research, may lead to quantitative estimates permitting incorporation of social parameters into "generic models of individual-level processes" (Martin et al. 2013). However, the discussion in Chap. 6, and so far, in this chapter does not provide much support for ecological factors as the "glue" for group maintenance, or for conditions that might favor the evolution of sociality; however, significant (?) energy savings and/or the potential or actual benefits derived from associating with kin are likely to provide the "glue."

7.3   A Case Study of Community Assembly: Superior and Inferior Competitors

Jones (1995) recorded interspecific interactions between members of 1 polygynandrous mantled howler group and 27 other taxa (Table 7.1). Negative interactions (interference competition) were most likely to occur when howlers and insects or birds utilized flowering trees. Fruit, on the other hand, appeared to facilitate mutualism and coexistence (e.g., mixed-species flocks). These observations showed that howlers were competitively inferior to members of some taxa (e.g., insects) on some spatiotemporally limited resources, interactions with the potential to keep howler numbers in check where costs from subordinate status translate into decreased feeding rates deleterious to survival or to relative reproductive success (Schoener 1971, 1974, 1982; DeJong 1976; Jones 2012). Under selection induced by interspecific competition, a species experiencing the lowest relative fitness outcomes (the ecologically "inferior" species, R^*), may be excluded temporarily or permanently (e.g., local extirpation) by the competitively superior species ("competitive exclusion principle").

Further studies are required to determine the generality of these findings. Possibly, insects and birds impose "ecological limitations" on mammals in many conditions, impacting effective population sizes, growth rates, population structure, including social organization of mammalian competitors' populations. Insects and birds, thus, may be major players in the assembly of mammalian communities, particularly, mutualisms among herbivores. The negative interactions displayed in Table 7.1 were most likely to occur when howlers and insects or birds utilized flowering trees; however, temporal partitioning was evident between bees and howlers, one of Chesson's (2000) proposed mechanisms for managing competition and a sign of facilitation by one or both taxa. Fruit appears to facilitate mutualism and coexistence. In order to explore the complete range of causes, mechanisms, and effects associated with the management of competition, social biologists must borrow concepts from Community Ecology.

Caveats are in order. For a resource to be utilized, it must be divisible by one or more of Chesson's (2000) dimensions (space, time, resource partitioning, "natural enemies"), lest competitive exclusion occur. For example, mantled howlers and *Centris* (?) bees both utilize *Andira inermis* flowers, monkeys for the whole flower (William Haber, personal communication), bees for pollen and nectar (Jones 2005). If howlers arrived at the tree before bees or if the apparent sensory distortions caused by the insects did not inhibit howlers from feeding, all of the flower parts would be consumed, extirpating or seriously depleting a valuable food resource from the bees' perspective. In the case tabulated, the large group of monkeys waited for up to 1 h until swarms of bees had perceptibly decreased in density and rate of movement (virulence; C. B. Jones, personal observation). Presumably, then, foraging on these flowers ceased to be economical for *Centris*, mediated by temporal partitioning of the apparently limiting resource, one of Chesson's (2000) proposed mechanisms of coexistence. In this case, howlers exhibited "restraint," possibly, a precursor to or a consequence of group formation and a factor influencing maintenance of groups.

Table 7.1 Summary of notes on interactions between mantled howler monkeys and 27 contraspecifics in Costa Rican tropical dry forest environment. (Jones 1995)

Class, genus, or species, and common names	Focal tree species, where identified, and notes
Insecta (15 %)	
Centris aethyctera, anthophorid bees	*Andira inermis* in flower producing nectar, bees interfere with howler feeding, howlers delay feeding until after diurnal pollination peak, bees displace monkeys (**competitive exclusion**)
Xylocopa spp., carpenter bees	*Gliricidia sepium* in flower, bees decrease average feeding rate of monkeys, bees interfere with howler feeding (**interference competition**)
Reptilia (7 %)	
Iguana iguana, Ctenosaura similis, iguanas	Feeding on fruit in *Licania arborea, Spondias* spp., *Ficus ovalis, Enterolobium cyclocarpum*, or *Cordia alliodora*, interspecific feeding associations (neutral or positive associations such as coexistence, facilitation, mutualism, synergy)
Aves (67 %)	
Cathartes aura, Caracara plancus, vultures	Female howlers emit appeasement calls to vultures, vultures displace young and adult female monkeys (**interference competition**)
Buteo magnirostris, Spizastur melanoleucus, hawks	Hawks displace howlers and some birds (e.g., jays) from feeding sites (**interference competition**)
Herpetotheres cachinnans, falcons	Falcons interfere with howler feeding, monkeys vocalize to falcons (**interference competition**)
Jabiru mycteria, storks	A low-flying stork triggers coordinated howls among adult male monkeys, storks interfere with howler feeding (**interference competition**)
Brotogeris jugularis, Aratinga canicularis, parrots	In fruiting tree, interspecific feeding associations (neutral or positive associations such as coexistence, facilitation, mutualism, synergy)
Eugenes fulgens, hummingbird	*Tabebuia neochrysantha* in flower, interspecific feeding association (neutral or positive associations such as coexistence, facilitation, mutualism, synergy)
Trogon spp., trogons	In fruiting tree, interspecific feeding associations (neutral or positive associations such as coexistence, facilitation, mutualism, synergy)
Eumomota superciliosa, Momotus lessonii, motmots	*Simarouba glauca* in fruit, birds pick fruit then leave tree to feed, motmots avoid howlers (**interference competition**)
Ramphastos spp., toucans	*Ficus ovalis* in fruit, mutual interference during feeding (**interference competition**)

Table 7.1 (continued)

Class, genus, or species, and common names	Focal tree species, where identified, and notes
Campephilus guatemalensis, Dryocopus lineatus, woodpeckers	Bird calls sound like howler barks, howlers may flush insects eaten by woodpeckers, apparent competition for space, birds may displace monkeys (**competitive exclusion?**)
Chiroxiphia linearis, manikins	Interspecific feeding in fruiting tree? (neutral or positive associations such as coexistence, facilitation, mutualism, synergy)
Cyanocorax spp., crows, jays	*Andira inermis, Anacardium excelsum, Muntingia calabura*, howlers may displace jays, howlers may displace insects (**interference competition**)
Campylorhynchus rufinucha, acacia wrens	*Simarouba glauca* in fruit, howlers flush insects, interspecific feeding association (positive association, apparently, facilitation)
Mammalia (11 %)	
Coendou mexicanum, Dasyprocta punctata, rodents	Howlers feeding on fruiting *Anacardium excelsum*, commensals beneath feeding tree (positive association)
Sciurus spp., squirrels	Fruiting *Ficus ovalis*, howlers displace squirrels, interspecific feeding association among howlers, ctenosaurs, parrots, trogons, jays, and squirrels (interference competition, apparently in context of neutral or positive associations)

Most of these events occurred when howlers and one or more additional species were feeding or attempting to feed on new leaves, fruit, or flowers, howlers' preferred food items, available primarily during dry season, November through April. Statistical analyses of these results (Jones 1995) revealed that interspecific interactions occurred more frequently in riparian (wetter) than deciduous (drier) habitat during dry season when clumped, ephemeral plant tissues of high quality favor mutual exploitation of food by guilds. Gilman et al. (2012) demonstrated theoretically that in any exploiter–victim association, the victim (e.g., a model such as an estrus female) can "win" the coevolutionary contest wherever the latter is able to lower its "interaction probabilities" with the exploiter (*via* mimicry (Fig. 8.1), fossorial habits, or aggregation). Where small population size yields low rates of interaction between victim and exploiter, effective population size of exploiters may decrease to extinction. See text and Jones (1995) for further discussion
Percentages = % total sample of 27 genera or species; type of interaction in parenthesis; negative interactions in bold

Consistent with the stress-gradient hypothesis and with formulations in behavioral ecology, the energetic demands imposed by a folivorous diet, such as the diet of mantled howlers, may favor group maintenance. Duffy (2002) discussed "the consumer connection" between community assemblages and ecosystem dynamics, and indeed held that in some biomes, including terrestrial ecosystems (Fig. 1.1), the actions of consumers are "equally and sometimes *more* likely to be manifested" than the action of primary producers, particularly, where top-down regulation dominates (italics added). On the other hand, Nee and May (1992), complementing the research findings of Gilman et al. (2012), provided theoretical models demonstrating that competitively inferior species may increase in number under certain environmental regimes, for example, where their dispersion is spatiotemporally "even" relative to their competitors. This finding has important implications for social mammals utilizing stressful mature ("old") leaves (e.g., "age-graded" and polygynandrous ateline and colobine monkeys, polygynous and "age-graded" gorillas, many rodents and ungulates), evenly dispersed food items relative to new leaves, flowers, and fruits. In sum, stress induced by differential interaction rates and low-quality food associated with increasing population density or resource competition may favor alternative allocation strategies, including facilitation and sociality, depending on their differential fitness costs and benefits to victims and exploiters, relatives and nonkin.

References

Agrawal AA, Klein CN (2000) What omnivores eat: direct effects of induced plant resistance on herbivores and indirect consequences for diet selection by omnivores. J Anim Ecol 69:525–535

Arthur SM, Prugh LR (2010) Predator-mediated indirect effects of snowshoe hares on Dall's sheep in Alaska. J Wild Manag 74:1709–1721

Barja I (2009) Decision-making in plant selection during the faecal-marking behaviour of wild wolves. Anim Behav 77:489–493

Bateman AW, Ozgul A, Nielsen JF, Coulson T, Clutton-Brock TH (2013) Social structure mediates environmental effects on group size in an obligate cooperative breeder, *Suricata suricatta.* Ecology 94:587–597

Bicca-Marques JC, Garber PA (2003) Experimental field study of the relative costs and benefits to wild tamarins (*Saguinus imperator* and *S. fuscicollis*) of exploiting contestable food patches as single- and mixed-species troops. Am J Primatol 60:139–153

Blumstein DT, Ebensperger LA, Hayes LD, Vásquez RA, Ahern TH, Burger JR, Dolezal AG et al (2010) Towards an integrated understanding of social behavior: new models and new opportunities. Front Behav Neurosci 4:1–9

Bourke AFG (2011). Principles of social evolution. Oxford University Press, Oxford

Bradbury JW, Vehrencamp SL (1977) Social organization and foraging in emballonurid bats III: mating systems. Behav Ecol Sociobiol 2:1–17

Carnicer J, Brotons L, Stefanescu C, Peñuelas J (2012) Biogeography of species richness gradients: linking adaptive traits, demography and diversification. Biol Rev 87:457–479

Chesson P (2000) Mechanisms of maintenance of species diversity. Ann Rev Ecol Syst 31:343–366

Chesson P, Kuang JJ (2008) The interaction between predation and competition. Nature 456: 235–238

Cody ML (1974) Competition and the structure of bird communities. Princeton University Press, Princeton

Cornulier T, Yoccoz NG, Bretagnolle V, Brommer JE, Butet A, Ecke F, Elston DA, Framstad E, Henttonen H, Hörnfeldt B, Huitu O, Imholt C, Ims RA, Jacob J, Jędrzejewska B, Millon A, Petty SJ, Pietiäinen H, Tkadlec E, Zub K, Lambin X (2013) Europe-wide dampening of population cycles in keystone herbivores. Science 340:63–66

Crook JH (1964) The evolution of social organization and visual communication in the weaver birds (Ploceinae). Brill, Leiden (Behaviour Supplement X)

Crook JH (1965). The adaptive significance of avian social organization. Symp Zool Soc Lond 14:181–218

Crook JH (ed) (1970) Social behaviour in birds and mammals. Academic, London

Daleo P, Iribarne O (2009) Beyond competition: the stress-gradient hypothesis tested in plant-herbivore interactions. Ecology 90:2368–2374

De Jager NR, Pastor J (2008) Effects of moose *Alces alces* population density and site productivity on the canopy geometries of birch *Betula pubescens* and *B. pendula* and Scots pine *Pinus sylvestris*. Wild Biol 14:251–262

De Jager NR, Pastor J, Hodgson AL (2009) Scaling the effects of moose browsing on forage distribution, from the geometry of plant canopies to landscapes. Ecol Monog 79:281–297

DeJong G (1976) Selection always increases efficiency. Am Nat 110:1013–1027

Dell AI, Pawar S, Savage VM (2013) Temperature dependence of trophic interactions are driven by asymmetry of species responses and foraging strategy. J Anim Ecol. doi:10.1111/1365-2656.12081

Duffy JE (2002) Biodiversity and ecosystem function: the consumer connection. Oikos 99:201–219

Emlen ST (1982). The evolution of helping. I. an ecological constraints model. Am Nat 119:29–39

Emlen ST, Oring LW (1977) Ecology, sexual selection, and the evolution of mating systems. Science 197:215–223

Everett A, Tong X, Briscoe AD, Monteiro A (2012) Phenotypic plasticity in opsin expression in a butterfly compound eye complements sex-role reversal. BMC Evol Biol 12:232

Fleming PJS, Allen BL, Ballard G-A (2012) Seven considerations about dingoes as biodiversity engineers: the socioecological niches of dogs in Australia. Austral Mammal 34:119–131

Fortin D, Morris DW, McLoughlin PD (2008) Habitat selection and the evolution of specialists in heterogeneous environments. Il J Ecol Evol 54:311–328

Gilman RT, Nuismer SL, Jhwueng D-C (2012) Coevolution in multidimensional trait space favours escape from parasites and pathogens. Nature 483:328–330

Givnish TJ (1999) On the causes of gradients in tropical tree diversity. J Ecol 87:193–210

Goheen JR, Palmer TM, Keesing F, Riginos C, Young TP (2010) Large herbivores facilitate savanna tree establishment *via* diverse and indirect pathways. J Anim Ecol 79:372–382

Holmgren M, Scheffer M (2010) Strong facilitation in mild environments: the stress gradient hypothesis revisited. J Ecol 98:1269–1275

Holzwarth FM, Kahl A, Bauhus J, Wirth C (2012) Many ways to die-partitioning tree mortality dynamics in a near-natural mixed deciduous forest. J Ecol. doi:10.1111/1365-2745.12015

Jasmin J-N, Zeyl C (2012) Life-history evolution and density-dependent growth in experimental populations of yeast. Evolution 66:3789–3802

Jennings AP, Veron G (2011) Predicted distributions and ecological niches of eight civet and mongoose species in Southeast Asia. J Mammal 92:316–327

Jones CB (1985) Reproductive patterns in mantled howler monkeys: estrus, mate choice, and copulation. Primates 26:130–142

Jones CB (1995) The potential for metacommunity effects upon howler monkeys. Neotrop Primates 3:43–45

Jones CB (2005) Discriminative feeding on legumes by mantled howler monkeys (*Alouatta palliata*) may select for persistence. Neotrop Primates 13:3–5

Jones CB (2012) Robustness, plasticity, and evolvability in mammals: a thermal niche approach. Springer, New York

Kawai T, Tokeshi M (2007) Testing the facilitation-competition paradigm under the stress-gradient hypothesis: decoupling multiple stress factors. Proc Roy Soc Lond B 274:2503–2508

Kearney MR, White CR (2012) Testing metabolic theories. Am Nat 180:546–565

Kerimoglu O, Straile D, Peters F (2013) Seasonal, inter-annual, and long-term variation in top-down vs. bottom-up regulation of primary production. Oikos 122:223–234

Kissui BM, Packer C (2004) Top-down population regulation of a top predator: lions in the Ngorongoro Crater. Proc Roy Soc Lond B 271:1867–1874

Lande R (2007) Expected relative fitness and the adaptive topography of fluctuating selection. Evolution 61:1835–1846

le Roux PC, McGeoch MA (2010) Interaction intensity and importance along two stress gradients: adding shape to the stress-gradient hypothesis. Oecologia 162:733–745

Letnic M, Ritchie EG, Dickman CR (2011) Top predators as biodiversity regulators: the dingo *Canis lupus dingo* as a case study. Biol Rev. doi:10.1111/j.1469-185X.2011.00203.x

Malkinson D, Tielbörger K (2010) What does the stress-gradient hypothesis predict? Resolving the discrepancies Oikos 119:1546–1552

Marcy AE, Fendorf S, Patton JL, Hadley EA (2013) Morphological adaptations for digging and climate-impacted soil properties define pocket gopher (*Thomomys* spp.) distributions. PLoS ONE 8:e64935

Marshall KN, Hobbs NT, Cooper DJ (2013) Stream hydrology limits recovery of riparian ecosystems after wolf reintroduction. Proc Roy Soc Lond B 280. doi:10.1098/rspb.2012.2977

Martin TE, Ton R, Nikilson A (2013) Intrinsic vs. extrinsic influences on life history expression: metabolism and parentally induced temperature influences on embryo development rate. Ecol Lett 16:738–745

McNaughton SJ, Ruess RW, Seagle SW (1988) Large mammals and process dynamics in African ecosystems. BioScience 38:794–800

Mielke HW (1977) Mound-building by pocket gophers (Geomyidae): their impact on soils and vegetation in North America. J Biogeog 4:171–180

Moreau G, Eveleigh ES, Lucarotti CJ, Quiring DT (2006) Ecosystem alteration modifies the relative strengths of bottom-up and top-down forces in a herbivore population. J Anim Ecol 75:853–861

Murray BD, Webster CR, Bump JK (2013) Broadening the ecological context of ungulate-ecosystem interactions: the importance of space, seasonality, and nitrogen. Ecology 94:1317–1326

Narwani A, Mazumder A (2013) Bottom-up effects of species diversity on the functioning and stability of food webs. J Anim Ecol. doi:10.1111/j.1365-2656.2011.01949.x

Nee S, May RM (1992) Dynamics of metapopulations: habitat destruction and comparative coexistence. J Anim Ecol 61:37–40

Prevedello JA, Dickman CR, Vieira MV, Vieira EM (2013) Population responses of small mammals to food supply and predators: a global meta-analysis. J Anim Ecol. doi:10.1111/1365-2656.12072

Prugh LR (2005) Coyote prey selection and community stability during a decline in food supply. Oikos 110:253–264

Prugh LR, Brashares JS (2011) Partitioning the effects of an ecosystem engineer: kangaroo rats control community structure via multiple pathways. J Anim Ecol 81:667–678

Queenborough SA, Metz MR, Wiegand T, Valencia R (2012) Palms, peccaries, and perturbations: widespread effects of small-scale disturbance in tropical forests. BMC Ecol 12:3. doi:10.1186/1472-6785-12-3

Rebollo S, Milchunas DG, Stapp P, Augustine DJ, Derner JD (2013) Disproportionate effects of non-colonial small herbivores on structure and diversity of grassland dominated by large herbivores. Oikos. doi:10.1111/j.1600-0706.2013.00403.x

Ricklefs RE (1977) On the evolution of reproductive strategies in birds: reproductive effort. Am Nat 111:453–478

Salgado-Luarte C, Gianoli E (2012) Herbivores modify selection on plant functional traits in a temperate rainforest understory. Am Nat 180:E42–E53

Sankaran M, Augustine DJ, Ratman J (2013) Native ungulates of diverse body sizes collectively regulate long-term woody plant demography and structure of a semi-arid savanna. J Ecol. doi:10.1111/1365-2745.12147

Schoener TW (1971). Theory of feeding strategies. Ann Rev Ecol Syst 2:369–404

Schoener TW (1974) Resource partitioning in ecological communities. Science 185:27–39

Schoener TW (1982) The controversy over interspecific competition: despite spirited criticism, competition continues to occupy a major domain in ecological thought. Am Sci 70:586–595

Schneider FD, Brose U (2012) Beyond diversity: how nested predator effects control ecosystem functions. J Anim Ecol. doi:10.1111/1365-2656.12010

Schnitzer SA, Carson WP (2001) Treefall gaps and the maintenance of species diversity in a tropical forest. Ecology 82:913–919

Sibley RM, Hone J (2003) Population growth rate and its determinants: an overview. In: Sibley RM, Hone J, Clutton-Brock TH (eds) Wildlife population growth rates. Cambridge University Press, Cambridge

Slobodkin LB (1980) Growth and regulation of animal populations. 2nd edn. Dover, New York

Sterck EHM, Watts DP, van Schaik CP (1997) The evolution of female social relationships in nonhuman primates. Behav Ecol Sociobiol 4:291–309

Sterck F, Markesteijn L, Schieving F, Poorter L (2011) Functional traits determine trade-offs and niches in a tropical forest community. Proc Natl Acad Sci U S A 108:20627–20632

Stien A, Ims RA, Albon SD, Fuglei E, Irvine RJ, Ropstad E, Halvorsen O, Langvatn R, Loe LE, Veiberg V, Yoccoz NG (2012) Congruent responses to weather variability in high Arctic herbivores. Biol Lett. doi:10.1098/rsbl.2012.0764

Takahashi K, Takahashi K (2013) Spatial distribution and size of small canopy gaps created by Japanese black bears: estimating gap size using dropped branch measurements. BMC Ecol 13. doi:10.1186/1472-6785-13-23

Teichman KJ, Nielsen SE, Rodand J (2013) Trophic cascades: linking ungulates to shrub-dependent birds and butterflies. J Anim Ecol. doi:10.1111/1365-2636.12094

Trainor BC, Crean KK, Fry WHD, Sweeney C (2010) Activity of extracellular signal-regulated kinases in social behavior circuits during resident-intruder aggression tests. Neuroscience 165:325–336

Wrangham RW (1980) An ecological model of female-bonded primate groups. Behaviour 75: 262–300

Chapter 8
The Evolution of Mammalian Sociality by Sexual Selection

Where populations have not evolved signals permitting interindividual proximity without a high likelihood of aggression, social evolution may be constrained.

Otte (1974)

Abstract This chapter addresses the evolution of mammalian sociality by sexual selection ("intrasexual" and "intersexual" selection), a topic directly related to the actions available to types maximizing "inclusive fitness." As females are "energy-maximizers," spatiotemporal distributions of females may entail significant (relative) fitness costs that males, "time-minimizers," may not be in a position to afford. Energy-saving strategies are also indicated for female mammals due to their high "reproductive load" and vulnerability to the effects of offspring competition. Energy savings is a thermal regulatory process defining natural and sexual selection, maintaining usable heat within limits propitious to optimal functioning (maintenance, growth, survival, reproduction). Thus, female traits may drive the evolution of male traits. Studies using *Drosophila melanogaster* as subjects showed that "sexual conflict" arises because "promiscuity" is incompatible with mutual interests, a finding with direct import for the evolution of mammalian sociality since promiscuity may have been the initial state of tolerance from which mammalian sociality evolved, linking sexual selection to the evolution of sociosexual assemblages in the class.

Keywords Sexual selection · Intersexual competition · Intrasexual competition · Sexual conflict · Eco-Ethology · Signaling · Male to female aggression · Social parasitism

Readers are referred to Andersson (1994), Trivers (1972), Otte (1979), Pizzari and Bonduriansky (2010), Westneat and Fox (2010, Sect. V), Davies et al. (2012), and Darwin (2004) for comprehensive reading on sexual selection theory, including extensions of it and recent advances in that field. In brief, Darwin (2004) recognized that some animal characteristics appeared to compromise rather than promote survival, thereby appearing to function counter to natural selection. Reasoning that traits deleterious to survival might be favored if sufficient benefits accrued to reproduction, Darwin proposed two mechanisms of "sexual selection," "intrasexual" and

C. B. Jones, *The Evolution of Mammalian Sociality in an Ecological Perspective,*
SpringerBriefs in Ecology, DOI 10.1007/978-3-319-03931-2_8,
© Clara B. Jones 2014

"intersexual" selection, driven by same-sex and between-sex competition, respectively (Darwin 2004; this brief, Sect. 3.3). West-Eberhard (1979; see Crook 1972) noted that both mechanisms entail intersexual competition for mates. Darwin (2004) was impressed that exaggerated structures (e.g., horns, antlers, colorful features) employed as sexual signals and displays were likely to expose types to predation, and these same, and other traits, may be costly to survival by increasing a type's vulnerability to parasitism, including "social parasitism" ("natural enemies"; Chesson 2000; this brief, Sect. 8.4), and increased intra- and interspecific competition.

The evolution of mating systems by sexual selection is well established (Crook 1972; Emlen and Oring 1977; Clutton-Brock and Harvey 1978; Clutton-Brock 1989; Davies et al. 2012), but intra- and intersexual competition as drivers of social evolution have received less focus in the mammalian literature (but see Nelson et al 2013; Clutton-Brock et al. 2006; and, for birds, Cornwallis et al. 2010). It is assumed herein that same-sex competition for mates and "mate choice" would follow the same rules as competition for other resources (nutrients, space) whereby facilitation of a conspecific's reproduction may represent a type's optimal response (Chap. 2) following Hamilton's rule. Traits of rivals and mates will vary, yielding types differentially successful in combat, display, and fertilization (males) or implantation (females). The intensity of competition for mates is expected to increase with increased population density and, under some conditions, reproductive groups of one or both sex may be favored. Following the scenarios addressed in Chap. 6 of this brief, groups may form as a result of interindividual interactions within populations "mapped" onto clumped resources, including mates, with the dynamics of those interactions bounded by the parameters of Hamilton's rule (Chaps. 1 and 2). Where types vary in sexual traits and where the spatiotemporal dispersion of mates is clumped, sexual selection may favor sociality within and/or between sexes, and associated traits may become "fixed" in a population.

8.1 The Energetics of Sexual Allocation

Schoener (1971; see Gittleman and Thompson 1988; Bergman et al. 2001) demonstrated theoretically that males are expected to be time-minimizers and females, energy-maximizers (see Sect. 3.1). These different life-history strategies are biased by initial reproductive allocations or energetic investments (Selman et al. 2012; Trivers 1972; see Proulx 1999). Queller and Strassmann (2010) pointed out that, compared to female insects, vertebrate females, particularly birds and mammals, invest heavily in reproduction, showing that taxonomic differences obtain (also see Selman et al. 2012; this brief, Sect. 5.4, Synopsis). The latter states may represent a "life insurance" strategy for one or both parents to protect reproductive investment in heterogenous regimes. Female mammals, as well, allocate nutrients to secondary sexual characteristics, such as pendulous breasts and other fatty deposits, presumably as characters facilitating mate attraction (see Jones 2007). Costly allocation of energy to reproduction and mating suggests a trade-off between efficiency and flexibility for female mammals (but see Trebatická et al. 2007; this brief, Chap. 2). Unlike most birds, characterized by biparental care, resource dispersion associated with most mammals' regimes has promoted "sexual segregation" and polygyny, apparently derived from promiscuous aggregations, as the primary sociosexual structures (Chap. 3). Apparently, the latter scenarios, combined with benefits of sexual segregation for males (a time-minimizing strategy?) favored female types predisposed to assume primary and obligate care of offspring. Female mammals, then, may have been under severe selective pressures to conserve resources (Clutton-Brock et al. 1989; Nagy et al. 1999), leading to the adoption of energy-conserving features associated with some social tactics and strategies (e.g., "coyness," selectivity, mate choice, "allomothering").

The aforementioned conditions and patterns are expected to enhance the basic asymmetry of male and female reproductive optima whereby females benefit most from control of fertilization, males from control of insemination (Alexander et al. 1997) The latter character state ("social selection") is predisposed to favor the evolution of signaling as an equalizing mechanism. Given the energetic constraints attendant to the female mammal, energy-saving counterstrategies to their fundamental reproductive conflict with potential mates ("sexual conflict"; Chapman et al. 2003), judicious mate "choice" and mate competition mechanisms ("intersexual" selection; see West-Eberhard 1979; Jones 2003), are bound to be critical counterstrategies to male persuasion, coercion, force, and control ("parasitism" by males; Davies et al. 2012). It may not be hyperbole to suggest that female mammals can ill afford to make a mistake in their choice of mates, predisposing them to select males with extreme (genetically correlated) traits, inducing positive feedback loops (Fisherian "runaway selection," Sect. 8.6). Schoener's (1971) and Trivers' (1972) formulations, then, encompass all conditions in which female mammals make "decisions" regarding reproductive allocation, including decisions to join (group formation) or remain in groups, linking natural selection to sexual selection and the evolution of mammalian sociality. A *caveat* to the study of reproductive strategies in female mammals must be evolution in heterogeneous regimes, conditions likely to have stressed "fitness budgets" as well as energy allocation tactics and strategies by increasing margins of error via decreasing likelihoods of accuracy.

Where female dispersion is determined by food dispersion, and if male dispersion is "mapped" onto dispersion of females (see Proulx 1999; this brief Chap. 6), male and female mammals do not, per se, compete for food but over mate "choice" (mate selectivity, "intersexual competition"). Males will exhibit mate "choice" to the degree that, for one reason or another, potential mates are unavailable (e.g., due to kinship, lack of female receptivity, lack or failure of male attraction, "social parasitism" or other forms of exploitation). Females are likely to exhibit female–female competition ("intrasexual selection") where attractive, available males are in short supply. Mechanisms of mate choice might have favored the evolution of mammalian social actions as mechanisms to manage competition for resources or for mates, conditions more likely to arise for energy-stressed females (female types with high l^* values relative to other female group members).

As polygyny and "sexual segregation" are the norms among mammals (Chaps. 3, 4 and 5), males in most populations are obligated to search for females, a response that may be costly in time that, at the extreme, may have favored male–female cores-idence (monogamy, "harems," or multimale–multifemale assemblies, Sects. 3.3 and 8.5). This strategy favors species recognition, high dispersal or colonization ability, and highly developed sensory perception (Ewer 1968; Wilson 1975). Since females are often spatiotemporally clumped in polygynous mammal species, promoting male search strategies, females may benefit from group living to facilitate their location by males (Sect. 8.6). On the other hand, in some mammalian taxa, females exhibit mate search strategies (e.g., in lekking species), and empirical studies demonstrate the high-energetic, including nutritional, expense of female mate search (pronghorns, *Antilocapra americana*, Byers et al. 2005, 2006; but see Trebatická et al. 2007). Mammalian taxa, such as pronghorns, in which females search for mates (lek-like), but where males do not display at a breeding site may represent an evolutionary precursor to "true" leks (Chap. 3). The costs of searching to females, combined with the already high "reproductive load" of female mammals, strongly suggest that, where this sex searches for mates, heterogeneous conditions render attractive, avail-able mates difficult to locate via increased spatiotemporal unpredictability. Where females search for mates as members of groups, they may experience an energy savings, this possibility as well as the previous topics are in need of systematic investigation.

As proposed, features associated with mate attraction may favor the evolution of female groups. A variety of signals and displays are ubiquitous among polygynous, including lekking, mammalian males, such as tusks, horns, antlers, colorful pelage, pendulous, large, or colorful testes and penises, and large body size (Ralls 1977). Female mammals may also exhibit structures to attract males, such as genital en-gorgement or exaggerated coloration to advertise fertility or receptivity (e.g., Jones 1985, 1997a). These signals and displays may function as appeasement to other females and may intensify male–male competition, facilitating mate assessment of male traits where females mate multiply. In addition, females of some species emit vocalizations before, during, and/or after assessment and/or copulation (rats, *Rattus norvegicus*; Thomas and Barfield 1985), audible responses that may attract social parasites (conspecifics) or predators ("natural enemies"; Chesson 2000), and that may heighten female–female competition for mates ("intrasexual competition," e.g., interference competition, "copying," "eavesdropping"). Sexual signals and displays observed in mammals will ultimately be defined by formal statements of "information theory" (Frank 2012; Proulx 2001).

The evolution of group living may benefit females if groups serve as "information centers" about female competitors and reproductive males in a population (Kerth and Reckardt 2003; Jerison 1983). Indeed, for these potential mates, benefits from advertisement may be highest where types cluster spatiotemporally (aggregations, Chaps. 3 and 6), "hotspots" beneficial to males as a time-minimizing strategy or to females as an energy maximization trajectory. For each sex, these different metabolic effects associated with "hotspots" are likely to decrease costs from mate search, broadcasting, and unpredictability as well as ignorance about competitors. Other features that may enhance attraction and proximity to the opposite sex are "nuptial feeding" and chemical marking ("urine ceremonies"; Schilder 1990) by males, and directed responses by females signaling changes in receptivity. Groups in a patch may vary in intensity of competition for mates and/or for competition for nutrients convertible to offspring, effects expected to influence relative benefits and costs to individuals from cooperation and/or altruism among kin on the one hand, and nonkin on the other. As mammalian females bear very high costs from allocation to reproduction (Clutton-Brock et al. 1989), this sex is expected to be most sensitive to variations in competitive regimes within and between patches.

8.2 "Sexual Conflict" Between Mammalian Males and Group-Living Females: Ecology Interacts with Traits

Where limiting resources and females are distributed unevenly, some males will control many more females than others, as found among most large mammals (Clutton-Brock 1989; this brief, Sect. 3). Accurate identification of mates is essential to each sex, and sexual selection has modified species identification and communication systems, acting differently on males and females (Clutton-Brock and Huchard 2013; Otte 1974, 1975). In most conditions, male mammals dominate females because: (1) body size of reproductive males is usually larger than that of reproductive females, (2) reproductive competition is more intense among males compared to females, (3) mammalian males living in groups are generally unrelated (e.g., in multimale–multifemale reproductive units), and (4) in the same conditions, males are generally able to increase their reproductive output more than females are able to. In other words, compared to females in the same patch, variance in reproductive success is expected to be higher in males (Trivers 1972). Additionally, females should prefer to control the timing of fertilization while males should prefer to control insemination (Alexander et al. 1997), creating conditions whereby different intersexual "fitness optima" reflect conflicts of interest, effects that should enhance likelihood of energy-stressed mammalian females exhibiting cooperation and/or altruism to others of their sex in a patch (Silk et al. 2003).

8.3 The Eco-Ethology of Male to Female Aggression

Among mammals, some environments have a high potential for male to female aggression (Estes 1992), reducing pressures on the differential fitness optima of each sex ("sexual conflict"; Rice 2000; Holland and Rice 1999; Chapman et al. 2003; Aloise King et al. 2013). Female mammals may be vulnerable to male persuasion, coercion, force, and parasitism, as well as to coordination and control by males, because high maternal investment predisposes females to phenotypes designed for efficient execution of maternal roles (physiological characteristics, mammaries, mate selectivity; Clutton-Brock et al. 1989). Although the ethological perspective holds that ritualized signals and displays function to decrease likelihood of aggression among conspecifics, the costs of producing these ritualized, nonstereotyped, or learned characteristics may significantly stress females' energy reserves (see Bondurianisky 2013). This condition pertains particularly to female mammals, obligated to and limited by extremely costly maternal allocation tactics and strategies (high "reproductive load").

For the aforementioned reasons, male and female mammals engage in an ongoing coevolutionary "arms race," imposing greater reproductive costs on each or "holding their own" in such a competitive "chase" (Chapman et al. 2003; Holland and Rice 1999). After Estes (1992), Ewer (1968), Clutton-Brock (1977), Chapman and Feldhamer (1982), Wasser (1983), Anderson and Jones (1984), and Mosser and Packer (2009), males appear to have won this race in some taxa (*Agouti*; Northern elephant seals, *Mirounga angustirostris*; walrus, *Odobenus rosmarus*; Hamadryas baboons, *Papio hamadryas*; chimpanzees, *Pan troglodytes*; lions, *Panthera leo*; domestic cats, *Felis catus*). In other taxa, females have apparently won (lemurs, Lemuridae; bonobo, *Pan paniscus*; mantled howler monkey, *Alouatta palliata*; coati, *Nasua narica*; African elephant, *Loxodonta africana*; reindeer, *Rangifer tarandus*), including species in which females are dominant to males. In a few species, intersexual relations have been characterized as "egalitarian" (striped mice, *Rhabdomys pumilio*; fox, *Lycaon*; muriquis, *Brachyteles aracnoides*), while in others, intersexual influence and "power" generally vary by context (Hawaiian monk seal, *Monachus schauinslandi*; squirrel monkeys, *Saimiri* spp.; most socially "monogamous" mammals involving single-male, single-female coresidence; "hierarchical" humans, *Homo sapiens*). The previous patterns may be influenced by alternative reproductive strategies employed by females and, particularly, males (P. C. Lee, personal communication; see Fig. 2.1).

Aggression, including coerced or forced copulation by males to females ("rape," "traumatic insemination," humans, Thornhill and Thornhill 1983; male orangutans, *Pongo pygmaeus*; see Brooks and Jennions 1999) is likely to be favored by selection where females of polygynous mammalian species ("harem," "age-graded"; red deer, *Cervus elaphus*; hartebeest, *Alcelaphus caama*) do not copulate outside their receptive period. Polygynous human systems represent one exception to this pattern. Where more than one female cycle concurrently in polygynous taxa and where females breed seasonally (pinnipeds), a female-biased "operational sex ratio" (OSR, relative occurrence of sexually active males to reproductive females) will obtain, a condition expected to (1) increase costs to males from attempts to monopolize cycling females, (2) restrict a polygynous male's temporal window for copulation and successful fertilization and, (3) generate intense male–male competition between polygynous males, between males governing relatively small and relatively large female groups, and between males without a group of females to coordinate and control.

In the aforementioned conditions, aggression by males may be favored if heritability (a population parameter measured as proportion of differences between types attributed to differences in alleles) reaches some threshold value relative to ecological and demographic factors, in particular, large and structured populations, as well as intensity of selection. The latter phenomena are directly related to genotype × environment interactions ("norms of reaction") and to intensities of competition within populations (within and between groups), for example, by variations in the effects of traits associated with male to female aggression. For example, in mammals, high levels of male to female aggression are associated with nonstereotyped (nonritualized) behavioral phenotypes (pinnipeds; ground squirrel, *Citellus armatus*; humans), high population density (pinnipeds; humans), breeding on land rather than in water (pinnipeds), a "catholic" (broad niche or opportunistic) diet (pinnipeds, humans), unstable male dominance hierarchies rather than resource or female defense (lions, Northern elephant seals, some bats), multiple mating by females (ubiquitous), polygynandry, "queuing" associated with hierarchical ("multilevel") group structures (humans, some cetaceans), and/or very lengthy periods of female pregnancy, lactation or maternal care (chimpanzees, humans). The previous tactics and strategies will promote or sustain a low r in groups, inhibiting the evolution of aggressive and reproductive restraint favoring the evolution of sociality. These conditions should favor, instead, intense within-group reproductive competition increasing likelihood of coexistence among unrelated types.

In a few taxa (*Agouti*, humans), social monogamy is associated with high levels of male agonism during courtship. Furthermore, male to female aggression is relatively common where females remain in their natal groups (most mammals), uncommon among patrilocal taxa (most atelids and apes, except chimpanzees with high rates of aggression (Wrangham and Peterson 1996) and humans exhibiting bisexual dispersal from natal groups (Hill et al. 2011). In some conditions, female dispersal may have been an adaptive counterstrategy to male coercion and force, an operation likely to maintain low r within groups if female dispersal is random by genotype. Among primates, for example, female dispersal is associated with energetically costly, more evenly dispersed, plant forage, particularly mature leaves, while matrilocal societies and male to female aggression are associated with nutritionally poor, clumped, ephemeral, fruit resources (but see anomalous spider monkeys, *Ateles*, and chimpanzees) .

In some cases (promiscuity "polyandry," "cryptic female choice") females are not readily monopolized by males (atelids, bonobos), decreasing effectiveness of male to female aggression, lowering the strength of sexual selection, and decreasing mean r in groups. Evolution of the latter female strategies are dependent upon the prior evolution of mechanisms for conflict management. Low levels of male to female aggression occur where females are dominant to males (Ralls 1976, 1977) and where females exert strong "choice" of mates ("leks"; polygynandry), suggesting the occurrence of effective mechanisms to manage or reduce "sexual conflict" or of spatial dispersions of limiting resources disfavoring agonistic interactions. On the other hand, sexual preferences by mammalian females may incur significant aggressive costs from nonpreferred males (Hamadryas baboons, Northern elephant seals). Additionally, mammalian males in several genera harass and coerce females with some frequency (*Halichoerus*, *Papio*), suggesting that phylogeny, in addition to ecology, needs to be considered as a correlate of male to female aggression promoting group life and associated traits. Regardless of the differential contributions of ecology and phylogeny to patterns of male to female aggression, a trade-off exists for both sexes whereby "female emancipation" (Jones and Cortés-Ortiz 1998) increases differences in *expressed*, *active*, and *effective* reproductive "optima" between the sexes.

Except for cases in which selection has preadapted reproductive mammalian males for subordinance to females (Ralls 1976), increasing the overlap of reproductive optima between the sexes, male to female aggression may have promoted the evolution of group life by enforcing female subordinance to males. This scenario may account, in part, for many examples of "sexual segregation" among mammals (Clutton-Brock et al. 1987) if "solitary" group structure emancipates reproductive females and males from direct negative consequences of coresidence occasioned by recurrent interindividual interactions driven by "sexual conflict." The analyses in this section suggest the testable proposition that, where mammalian males and females coreside, some stable degree of overlap in reproductive optima between the sexes must obtain resulting from shared kinship ("nested" bottlenose dolphins, *Tursiops*; Wiszniewski et al. 2010) or "shared reproductive fates" (cooperatively breeding kit foxes, *Vulpes macrotis mutica*; "nested" humans) .

8.4   A Simple Model of Male to Female Aggression in Mammals

Male to female aggression (persuasion coercion, force) may be modeled as male parasitism of a reproductive member of the opposite sex whereby a male exploits a female for reproductive advantage ("social parasitism"; Jones 1997b, 2005; Emerson 1958; Wheeler 1906; Wilson 1971; Michener 1974; Hölldobler and Wilson 1990), and Davies et al. (2012) showed that male (parasite) to female (host) parasitism is a sexually selected trait. "Social parasitism" may be considered one type of Chesson's (2000) category, "natural enemies," and the latter factors may favor facilitation as a means of conflict management in some conditions. Quantitative modeling puts social parasitism in perspective. Consider a male aggressor, the Sender, exploiting the time–energy budget of a reproductive female (a Receiver). Following May and Anderson (1990, in Moore 2002), Moore pointed out that fitness of a parasite (here, an adult male aggressor) can be measured as reproductive rate (R_0), a density-dependent value (Gill 1974). May and Anderson's equation formalizes virulence (rate of deleterious effects of male to female aggression) by way of a measure of cost to a female's fitness (increased intensity of intra- and intersexual interactions). May and Anderson's equation can be modified for male parasitism of females such that

$$R_0 = y(N)/(a + b + v),$$

where y is transmission rate (= "virulence," in the present case, reproductive costs imposed upon females by males), N is population density of reproductive females, a is rate of cost to reproductive females, b is rate of cost to reproductive females from all but virulence ("opportunity costs"), and v is a host's (a female Receiver's) recovery rate (a female's ability to completely or partially escape) from deleterious reproductive effects of a parasite's (aggressor male) responses (e.g., by increasing future reproductive rate or exploiting a mutation for an effective counterstrategy to male parasitism, such as increasing defensive networks with other females). May and Anderson's formula might be employed to predict conditions under which benefits from male parasitism decrease (e.g., where virulence, transmission, and recovery rate are independent; Moore 2002).

Females (hosts) may effect counterstrategies to male persuasion, coercion, and force, though males may control virulence at a level sufficient to coordinate and control hosts but not virulent enough to induce female counterstrategies ("immune response"). Such a state will limit a coevolutionary "chase" ("arms race") between the sexes (Rice 2000; Holland and Rice 1999), increasing the value to female hosts of associating in defensive networks (groups) with other females and with nonaggressive males, mitigating sexual conflict and its attendant costs. It is important to keep in mind that social parasitism, like other exploitative strategies, cannot induce altruism unless Hamilton's rule is satisfied.

Furthermore, if females are more canalized than males (Jones 1980), mammalian females may not display sufficient genetic heterogeneity to counter male parasitism, and increasing virulence will be costly to both sexes where female reproductive rate is significantly compromised via morbidity or mortality. Thus, in addition to affecting inclusive fitness of mates, male parasitism of females has the potential to affect growth rate of groups and mean fitness of populations, a condition that may increase or decrease intra- and interspecific competition with consequences for variations in intragroup levels of competition affecting differential benefits and costs from joining networks of kin or nonkin (West et al. 2002; De Bruyn 1980). Finally, the effectiveness of parasitic strategies by males may, in some conditions, depend upon the ability of reproductive females to discriminate parasitic from nonparasitic males, another topic requiring theoretical and empirical investigation. In sum, any increases in the reproductive load of females will depress their reproductive rates and outputs of offspring. This condition, beyond some threshold value for each type, could bias females to indirect reproduction (sociality).

Progress in these areas of research should be significantly promoted by Bourke's (2011) proposition that shared reproductive interests (e.g., between kin) and/or shared "reproductive fates" (e.g., between mates) should stabilize the evolution of groups and sociality, effects that may facilitate the evolution of cooperation and/or altruism among females as defensive and/or energy-maximizing strategies. Shared interests between mammalian reproductives may be forced upon prospective mates by male parasitism inducing a variety of subordinate traits in females (e.g., a large repertoire of submissive behaviors). Male parasitism of females via aggression may have originally favored increased maternal investment (i.e., costly gestation and lactation relative to body size), emancipating males by time savings and investments in male–male competition for monopolization of mates. For mammalian males, costly time minimization strategies may have been induced by environmental heterogeneity, the source of stressors (stimuli depressing reproductive rates) widely applicable to contexts in which mammals evolved (Jones 2009; Southwood 1977).

The possibility that male parasitism induced costly reproduction in mammalian females is suggestive of an evolutionary "arms race" between the sexes, though increased maternal investment by female mammals may have enhanced reproductive interests ("shared fates") between mates. Michener (1974) classified social parasites as "natural enemies," one of Chesson's (2000) mechanisms for the management of competition. The latter author employed "natural enemies" when discussing interspecific competition. However, Wilson's (1971) treatment shows that social parasitism is, as well, characteristic of intraspecific relations, in particular, closely associated types, including members of the same lineage. The foregoing ideas might be amenable to comparative tests, including mammalian females' vulnerability to parasitism from other sources (offspring, other females; see Lewis and Pusey 1997; Jones 2005; Galef 1991).

8.5   Managing Conflict Where More than One Males Coreside with Reproductive Females

More than one reproductive males cohabiting in stable groups with reproductive females are virtually limited to mammals (Wilson 1975; Brown 1975; this brief, Sects. 3.3, 3.5, Chap. 4), and most empirical reports of these structures remain descriptive rather than theoretical, hypothetico-deductive, or empirical, including experimental (but see Jones 1982). A paucity of studies is available to describe degrees of relatedness, intrasexual competition, or tendencies for these males to exhibit mate "choice." Additionally, systematic research on the stability of "fission–fusion" dynamics, frequently characterizing multimale–multifemale and "nested" reproductive groups, has not been conducted. In both multimale–multifemale and "nested" societies, males demonstrate hierarchies, coalitions, and alliances, but mammalian males rarely, if ever, demonstrate altruism, achievable only via shared genes among relatives.

Recent reports on polygynandrous lions (Mosser and Packer 2009) and hierarchically organized bottleneck dolphins (Wiszniewski et al. 2012) suggest that, in some conditions, defense of reproductive females may explain benefits to related or unrelated males. The latter reports indicated, as well, that (up to some limit) larger group sizes are associated with greater reproductive benefits to males (though not necessarily to females?). Discussing eusocial bathyergids, Lewis and Pusey (1997; also see Horwich et al. 2001) reported that higher infant mortality was associated with larger groups, a trend that, if common among mammals, would oppose Allee effects (Allee 1931) whereby female reproductive success increases with an increase in group size. Compared to sociality among females, the scientific literature on sociality among mammalian males is limited, a topic in need of systematic study, particularly, variations in tactics and strategies for the management of competition attendant to reproductive conflicts of interest as well as differential behaviors and network characteristics of related and unrelated reproductive males. Male dominance hierarchies are ubiquitous in multimale–multifemale assemblies, and a type's condition- and spatiotemporally dependent dominance rank should be decomposable into traits co-varying with fitness in fluctuating environments. In theory, these traits covary, as well, with measures of sexual selection (e.g., male–male displacements, copulation rates, rates of signaling and displaying). Controlled experiments in seminatural are-nas are needed to separate the effects of female traits on a male's traits, such as different fitness optima between the sexes, patterns of female choice, quality of care for a male's offspring, and "polyandry." Finally, male dominance hierarchies manage intrasexual competition among males whereby each type struggles for "maximum viability" and the relatively lowest levels of l^*_{within}.

8.6   A Final Note on Females: Potentials and Constraints

For mammalian females, energy savings drives the selection of traits (Schoener 1971; see Russell et al. 2003), a thermal regulatory process maintaining usable heat within limits propitious to optimal maintenance, survival, reproduction, and growth (Gittleman and Thompson 1988; McNab 1980). As females are "energy-maximizers," sexually selected signals and displays may represent a significant cost to inclusive fitness that, in the same conditions, males, "time-minimizers," may be in a better position to afford (see Clutton-Brock et al. 1989). Mammalian males can significantly influence population parameters by controlling reproductive careers of females. Such influence can be enhanced by ecological, by tactical and strategic decision-making (male herding behavior, infanticide, "sneaking"), or by females themselves (passive or "cryptic" "female choice," providing information to males about reproductive state, facilitation of male intromission, repelling adult female or juvenile interference). Whatever the precise environmental components determining the reproductive strategies of mammalian females, their life-history "decisions" are expected to be a function of life in (thermally) heterogeneous regimes (Geisel 1976; Schaffer 1974). In theory, female traits and environmental filters are measurable using taxon-independent criteria permitting quantitative analyses within and between species and within and between "patches" (Fig. 8.1).

There is a critical need to investigate the reaction norms of female mammals and the strength of selection pressures on female traits (e.g., intrasexual selection; Clutton-Brock et al. 2006). Qvarnström (2001) discussed reports showing that male traits attractive to females might vary in their effectiveness across changing con-ditions (spatial and temporal) and that females may gain reproductive advantages via tactics and strategies (genetically correlated traits) other than by favoring "good genes." These results using insects as subjects provide testable hypotheses for re-search projects targeting male and female strategies, and the extent to which male and female fitness optima vary by condition has received little attention in stud-ies of mammals (Sect. 5.3). Also, females often reside in groups in polygynous and polygynandrous as well as primitively eusocial mammals, condition-dependent structures reflecting tolerance and shared interests (e.g., energy savings) in partic-ular environmental regimes. The potential for patterns of female groupings, and

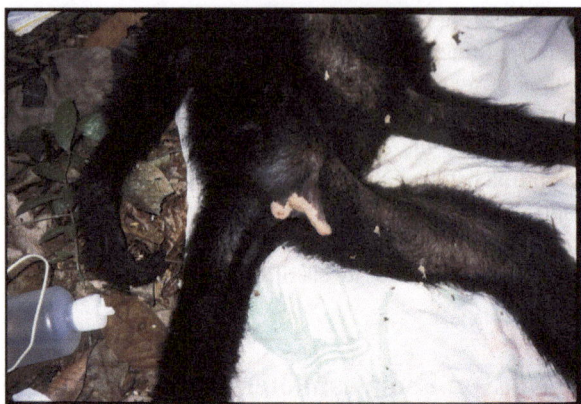

Fig. 8.1 Anesthetized adult female Peruvian ("black") spider monkey, *Ateles chamek* (Primates, Atelidae) exhibiting mimicry of male scrotum (also see wildebeest, *Connochaetes taurinus*), entailing a large, pendulous clitoris. This clitoris is an example of a "trait" that can be defined as "a physical, biochemical, morphological, physiological, phenological, or behavioral feature measurable at the individual level, from the cell to the whole-organism level" (Carnicer et al. 2012). When features of types are identified as (independent) standardized measurements, quantitative treatments can be conducted within and between taxa. This case of (aggressive? defensive?) morphological mimicry may represent an exaggerated (defensive? sexual? aggressive?) display favored by selection in response to intraspecific, intrasexual, or intersexual competition for food or other limiting resources (see Stankowich and Caro 2009). Across mammals, genital hypertrophy (e.g., female mantled howler monkeys) may be an evolutionary precursor to scrotal mimicry that may be ancestral to peniform, erectile clitorises (spotted hyena, *Crocuta crocuta*), and each state may induce "rapid" evolution of traits in other types (conspecific or contraspecific) affected by the display if "arms races" are operating. These displays may also function as species recognition devices (see Eibl-Eibesfeldt 1970). In mantled howler monkeys, also atelids, variations in vulval color, volume of vaginal excretions, and morphology may distinguish subspecies (C. B. Jones, personal observation; see Jones 1985, 1997a). This photograph was taken at Lago Caiman, Noel Kempff National Park, Bolivia, by © Rob Wallace

other female responses (e.g., allomaternal care, hygienic grooming, adoption, "interference" competition) to influence, if not manage, competitive relations among males is virtually unstudied; however, the extent to which females in polygynandrous assemblages exhibit "female emancipation" (Andersson 2005) is noteworthy. As Andersson's (2005) paper suggests, "female emancipation" may be identified wherever females mate multiply, a virtually ubiquitous trait of female mammals almost certainly retained from the ancestral "promiscuous" toolkit.

Holland and Rice (1999) removed effects of sexual selection in one experimental population, finding that males' virulent traits and their deleterious effects on females and on population growth rates "diminished" (sexual segregation or monogamy?) compared to a control population permitted to evolve with selection by sex unimpeded. Although toxicity of male sperm and female resistance decreased in the experimental condition in the previous study (see Gomendio and Roldan 1993), studying coevolution between "male ejaculates and female reproductive biology," showed that sexual selection may entail benefits as well as costs for mammalian females. A detailed understanding of intrinsic and extrinsic constraints on female life-history tactics and strategies requires further study in mammals. However, high "reproductive load" is expected to burden female mammals as a result of uncommonly high zygote and maternal allocation strategies (Trivers 1972), including attendant thermal requirements (Gittleman and Thompson 1988). It would seem that, for the aforementioned and other reasons, female mammals are predisposed to "social neglect," highlighting the advantages of closely monitoring ("record keeping") their interactions with members of their group, possibly, to gain benefits for the lowest possible cost. Females experiencing "social neglect," no doubt run the risk of a "social trap" whereby they may be, *ceteris paribus*, destined to one or more "helper" (dependent) roles.

References

Alexander RD, Marshall DC, Cooley JB (1997) Evolutionary perspectives on insect mating. In: Choe JC, Crespi BJ (eds) The evolution of mating systems in insects and arachnids. Cambridge University Press, Cambridge

Allee WC (1931) Animal aggregations, a study in general sociology. The University of Chicago Press, Chicago

Aloise King ED, Banks PB, Brooks RC (2013) Sexual conflict in mammals: consequences for mating systems and life history. Mamm Rev 43:47–58

Andersson M (1994) Sexual selection. Princeton University Press, Princeton

Andersson M (2005) Evolution of classical polyandry: three steps to female emancipation. Ethology 111:1–23

Anderson S, Jones JK Jr (eds) (1984) Orders and families of recent mammals of the world. Wiley-Interscience, New York

Bergman CM, Fryxell JM, Gates CC, Fortin D (2001) Ungulate foraging strategies: energy-maximizing or time-minimizing? J Anim Ecol 70:289–300

Bonduriansky R (2013) The ecology of sexual conflict: background mortality can modulate the effects of male manipulation on female fitness. Evolution. doi:10.1111/evo.12272

Bourke AFG (2011) Principles of social evolution. Oxford University Press, Oxford

Brooks R, Jennions MD (1999) The dark side of sexual selection. Trends Ecol Evol 14:336–337

Brown JL (1975) The evolution of behavior. W. W. Norton, New York

Byers JA, Wiseman PA, Jones L, Roffe TJ (2005) A large cost of female mate sampling in pronghorn. Am Nat 166:661–668

Byers JA, Byers AA, Dunn SJ (2006) A dry summer diminishes mate search effort by pronghorn females: evidence for a significant cost of mate-search. Ethology 112:74–80

Carnicer J, Brotons L, Stefanescu C, Peñuelas J (2012) Biogeography of species richness gradients: linking adaptive traits, demography, and diversification. Biol Rev 87:457–479

Chapman JA, Feldhamer GA (eds) (1982) Wild mammals of North America: biology, management, and economics. Johns Hopkins University Press, Baltimore

Chapman T, Arnqvist G, Bangham J, Rowe L (2003) Sexual conflict. Trends Ecol Evol 18:41–47

Chesson P (2000) Mechanisms of maintenance of species diversity. Ann Rev Ecol Syst 31:343–366

Clutton-Brock TH (ed) (1977) Primate ecology. Academic, New York

Clutton-Brock TH (1989) Mammalian mating systems. Proc Roy Soc Lond B 236:339–372

Clutton-Brock TH, Harvey PH (1978) Mammals, resources, and reproductive strategies. Nature 273:191–195

Clutton-Brock TH, Huchard E (2013) Social competition and its consequences in female mammals. J Zool 289:151–171

Clutton-Brock TH, Iason GR, Guiness FE (1987) Sexual segregation and density-related changes in habitat use in male and female red deer (*Cerrus elaphus*). J Zool 211:275–289

Clutton-Brock TH, Albon SD, Guiness FE (1989) Fitness costs of gestation and lactation in wild mammals. Nature 337:260–262

Clutton-Brock TH, Hodge SJ, Spong G, Russell AF, Jordan NR, Bennett NC, Sharpe LL, Manser MB (2006) Intrasexual competition and sexual selection in cooperative mammals. Nature 444: 1065–1068

Cornwallis CK, West SA, Davis KE, Griffin AS (2010) Promiscuity and the evolutionary transition to complex societies. Nature 466:969–972

Crook JH (1972) Sexual selection, dimorphism, and social organization in the primates. In: Campbell B (ed) Sexual selection and the descent of man, 1871–1971. Aldine, Chicago

Darwin C (2004, 1871) Descent of man and selection in relation to sex. Barnes & Noble, New York

Davies NB, Krebs JR, West SA (2012) An introduction to behavioural ecology, 4th edn. Wiley-Blackwell, Oxford

De Bruyn GJ (1980) Coexistence of competitors: a simulation model. Netherlands J Zool 30:345–368

Eibl-Eibesfeldt I (1970) Ethology: the biology of behavior. Holt, Rinehart, & Winston, New York

Emerson AE (1958) The evolution of behavior among social insects. In: Roe A, Simpson GG (eds) Behavior and evolution. Yale University Press, New Haven

Emlen ST, Oring LW (1977) Ecology, sexual selection, and the evolution of mating systems. Science 197:215–223

Estes R (1992) The behavior guide to African mammals including hoofed mammals, carnivores, and primates. The University of Chicago Press, Chicago

Ewer RF (1968) Ethology of mammals. Logos, London

Frank SA (2012) Natural selection. V. How to read the fundamental equations of evolutionary change in terms of information theory. J Evol Biol 25:2377–2396

Galef BJ (1991) Information centers of Norwegian rats: sites for information exchange and information parasitism. Anim Behav 41:295–301

Geisel JT (1976) Reproductive strategies as adaptations to life in temporally heterogeneous environments. Ann Rev Ecol Syst 7:7–80

Gill DE (1974) Intrinsic rate of increase, saturation density, and competitive ability. II. the evolution of competitive ability. Am Nat 108:103–116

Gittleman JL, Thompson SD (1988) Energy allocation in mammalian reproduction. Am Zool 28:863–875

Gomendio M, Roldan E (1993) Coevolution between male ejaculates and female reproductive biology in eutherian mammals. Proc Roy Soc Lond B 252:7–12

Hill KR, Walker RS, Božičvić M, Eder J, Headland T, Hewlett A, Hurtado AM, Marlowe F, Wiessner P, Wood B (2011) Co-residence patterns in hunter-gatherer societies show unique human social structure. Science 331:1286–1289

Holland B, Rice WR (1999) Experimental removal of sexual selection reverses intersexual antagonistic coevolution and removes a reproductive load. Proc Natl Acad Sci U S A 96:5083–5088

Hölldobler B, Wilson EO (1990) The ants. Harvard University Press, Cambridge

Horwich RH, Brockett RC, James RA, Jones CB (2001) Population structure and group productivity of the Belizean black howling monkey (*Alouatta pigra*): implications for female socioecology. Primate Rep 61:47–65

Jerison HJ (1983) The evolution of the mammalian brain as an information-processing system. In: Eisenberg JF, Kleiman DG (eds) Advances in the study of mammalian behavior. American Society of Mammalogists, Shippensburg

Jones CB (1982) A field manipulation of spatial relations among male mantled howler monkeys. Primates 23:130–134

Jones CB (1985) Reproductive patterns in mantled howler monkeys: estrus, mate choice, and copulation. Primates 26:130–142

Jones CB (1997a) Subspecific differences in vulva size between *Alouatta palliata palliata* and *A. p. mexicana*: implications for assessment of female receptivity. Neotrop Primates 5:46–48

Jones CB (1997b) Social parasitism in the mantled howler monkey, *Alouatta palliata* Gray, (Primates: Cebidae [now Atelidae]): involuntary altruism in a mammal? Sociobiology 30:51–61

Jones CB (2003) Sexual selection and reproductive competition in primates. American Society of Primatologists, Norman, OK

Jones CB (2005) Social parasitism in mammals with particular reference to Neotropical primates. Mastozoológica Neotropical 12:19–35

Jones CB (2007) Orgasm as a post-copulatory display. Arch Sex Behav 36:633–636

Jones CB (2009) The effects of heterogeneous regimes on reproductive skew in eutherian mammals. In: Hager R, Jones CB (eds) Reproductive skew in vertebrates: proximate and ultimate causes. Cambridge University Press, Cambridge

Jones CB (2012) Robustness, plasticity, and evolvability: a thermal niche approach. Springer, New York

Jones CB, Cortés-Ortiz L (1998) Facultative polyandry in the howling monkey (*Alouatta palliata*): Carpenter was correct. Bol Primatol Lat 7:1–7

Kerth G, Reckardt TK (2003) Information transfer at roosts in female Bechstein's bats. Proc Roy Acad Sci 270:511–515

Lewis SE, Pusey AE (1997) Factors influencing the occurrence of communal care in plural breeding mammals. In: Solomon NG, French JA (eds) Cooperative breeding in mammals. Cambridge University Press, New York

McNab BK (1980) Food habits, energetics, and the population biology of mammals. Am Nat 116:106–124

Michener CD (1974) The social behavior of the bees: a comparative study. Harvard University Press, Cambridge

Moore J (2002) Parasites and the behaviour of animals. Oxford University Press, Oxford

Mosser A, Packer C (2009) Group territoriality and the benefits of sociality in the African lion. Anim Behav 78:359–370

Nagy KA, Girard IA, Brown TK (1999) Energetics of free-ranging mammals, reptiles, and birds. Ann Rev Nutrition 19:91–122

Nelson AC, Colson KE, Harmon S, Potts WK (2013) Rapid adaptation to mammalian sociality via sexually selected traits. BMC Evol Biol 243:1–14

Otte D (1974) Effects and functions in the evolution of signaling systems. Ann Rev Ecol Syst 5:385–417

Otte D (1975) On the role of intraspecific deception. Am Nat 109:239–242

Otte D (1979) Historical development of sexual selection theory. In: Blum MS, Blum NA (eds) Sexual selection and reproductive competition in insects. Academic, New York

Pizzari T, Bonduriansky R (2010) Sexual behaviour: conflict, cooperation, and coevolution. In: Szekely T, Moore A, Komdeur J (eds) Social behaviour: genes, ecology, and evolution. Cambridge University Press, Cambridge

Proulx SR (1999) Mating systems and the evolution of niche breadth. Am Nat 154:89–98

Proulx SR (2001) Can behavioural constraints alter the stability of signaling equilibria? Proc Roy Soc Lond B 268:2307–2313

Queller DC, Strassmann JE (2010) Evolution of complex societies. In: Westneat DE, Fox CW (eds) Evolutionary behavioral ecology. Oxford University Press, Oxford

Qvarnström A (2001) Context-dependent genetic benefits from mate choice. Trends Ecol Evol 16:5–7

Ralls K (1976) Mammals in which females are larger than males. Q Rev Biol 51:245–276

Ralls K (1977) Sexual dimorphism in mammals: avian models and unanswered questions. Am Nat 111:917–938

Rice WR (2000) Dangerous liaisons. Proc Nat Acad Sci U S A 97:12953–12955

Russell AF, Sharpe LL, Brotherton PNM, Clutton-Brock TH (2003) Cost minimization by helpers in cooperative vertebrates. Proc Nat Acad Sci USA 100:3333–3338

Schaffer WM (1974) Optimal reproductive effort in fluctuating environments. Am Nat 108:783–790

Schilder MBH (1990) Interventions in a herd of semi-captive plains zebras. Behaviour 112:53–83

Schoener TW (1971) Theory of feeding strategies. Ann Rev Ecol Syst 2:369–404

Selman C, Blount JD, Nussey DH, Speakman JR (2012) Oxidative damage, ageing, and life-history evolution: where now? Trends Ecol Evol 27:570–577

Silk JB, Alberts SC, Altmann J (2003) Social bonds of female baboons enhance infant survival. Science 302:1231–1234

Southwood TRE (1977) Habitat, the templet for ecological strategies? J Anim Ecol 46:337–365

Stankowich T, Caro T (2009) Evolution of weaponry in female bovids. Proc Roy Soc Lond B 276:4329–4334

Thomas DA, Barfield RJ (1985) Ultrasonic vocalizations of the female rat (*Rattus norvegicus*) during mating. Anim Behav 33:720–725

Thornhill R, Thornhill NW (1983) Human rape: an evolutionary analysis. Ethol Sociobiol 4: 137–173

Trebatická L, Ketola T, Klemme I, Eccard JA, Ylönen H (2007) Is reproduction really costly? Energy metabolism of bank vole (*Clethrionomys glareolus*) females through the reproductive cycle. Ecoscience 14:306–313

Trivers RL (1972) Parental investment and sexual selection. In: Campbell B (ed) Sexual selection and the descent of man, 1871–1971. Aldine, New York

Wasser SK (ed) (1983) Social behavior of female vertebrates. Academic, New York

West SA, Pen I, Griffin AS (2002) Cooperation and competition between relatives. Science 296: 72–75

West-Eberhard MJ (1979) Sexual selection, social competition, and evolution. Proc Am Phil Soc 123:222–234

Westneat DE, Fox CW (eds) (2010) Evolutionary behavioral ecology. Oxford University Press, Oxford

Wheeler WM (1906) On the founding of colonies by queen ants, with special reference to the parasitic and slave making species. Bull Am Mus Nat Hist 22:33–105

Wilson EO (1971) The insect societies. Belknap, Cambridge

Wilson EO (1975) Sociobiology: the new synthesis. Belknap, Cambridge

Wiszniewski J, Lusseau D, Möller LM (2010) Female bisexual kinship ties maintain social cohesion in a dolphin network. Anim Behav 80:895–904

Wiszniewski J, Corrigan S, Beheregaray LB, Möller LM (2012) Male reproductive success increases with alliance size in Indo-Pacific bottlenose dolphins (*Tursiops aduncus*). J Anim Ecol 81: 423–431

Wrangham R, Peterson D (1996) Demonic males: apes and the origins of human violence. Houghton Mifflin Company, Boston

Chapter 9
Proximate Causation: Functional Traits and the Ubiquity of Signaler to Receiver Interactions: From Biochemical to Whole Organism Levels of Mammalian Social Organization

The molecular functions of many genes are highly conserved across species, even for complex traits.

Robinson et al. (2005)

Neurophysiological studies in the lab have revealed neural correlates of stimulus and movement value in parietal cortex and cingulated cortex, neural circuits implicated in attention, emotion, and decision-making.

Platt (2013)

Understanding of the behavioural mechanisms driving density-dependent processes provides potentially much greater insights than simply describing the population-level processes directly.

Sutherland and Norris (2003)

Abstract This chapter summarizes selected proximate correlates of mammalian sociality, including genetic, genomic, and physiological correlates. Ecological correlates were discussed in Chaps. 6 and 7. Mammalian sociogenomics (gene ontological studies) is in its early stages but has the potential to address questions concerning the play of phenotypes in fluctuating environments (Coda). The role of oxytocin and dopamine is discussed, including an overview of the new field "ecological neuroscience," conducting laboratory research on testing ecological theories (e.g., optimal foraging theory) as they pertain to physiological variables. This chapter also addresses "rapid evolution" as a process explaining mammalian sociality.

Keywords Functional traits · Social genetics · Sociogenomics · Social neuroscience · Socioecology · Social "toolkit" · Generalist phenotypes

Bounded by Hamilton's rule, genes, physiology, development, and morphology underlie the actions available to a type in the face of competition, in particular, the condition-dependent effects of intraspecific competition on a type's reproductive rate. Animal behavior may be conceptualized as one or more action pattern, motor pattern,

C. B. Jones, *The Evolution of Mammalian Sociality in an Ecological Perspective*, SpringerBriefs in Ecology, DOI 10.1007/978-3-319-03931-2_9,
© Clara B. Jones 2014

or performance capable of movement because of morphological, often, anatomical, components driven by neuromuscular events (Eibl-Eibesfeldt 1970, 2007). Behaviors are a component of an organism's "phenotype," the surface of a whole organism exposed directly to the abiotic (temperature, humidity, soil) and biotic (plants, predators, conspecifics) environments, including gradients. An action pattern incorporates the potential to communicate information and to serve as a signal whether or not it is genetically correlated (Proulx 2001), and a behavior can be used by a receiver "to alter the probability of interacting with particular signallers" (Proulx 2001). Most studies of the biochemistry, including genetics and genomics, physiology, and development of mammals are not conducted with ecology and evolutionary biology as a central focus, nor does data collection generally occur in natural conditions. Several studies indicated that social features are directly influenced by the actions of single genes and their pathways. Although, in none of these reports were pathways specified. Ferguson et al. (2000) showed that, unlike wild-type mice, mutant males of the same species, deficient for the oxytocin gene, "failed to develop social memory." In 2002, Stowers et al. (2002) found that male mice deficient in the ion channel, TRP2, lost the ability to discriminate sex and to express same-sex agonistic responses.

A highly publicized study by Lim et al. (2004; Wang et al. 2013) demonstrated "enhanced partner preference (among males) in a promiscuous (vole) species by manipulating the expression of a single gene." By transferring a gene expressing relatively higher levels of vasopressin from a monogamous vole species (*Microtus ochrogaster*) into the ventral forebrain of a promiscuous vole species (*M. pennsylvanicus*), these authors were able to induce more selective mate choice. In each case showing "social" behavior mediated by a single gene or single-gene product, effects pertain to male responses, supporting Lerner's (1954; also see Jones 2012) findings for *Drosophila melanogaster* that females are the more canalized sex (Chap. 8). Each of the aforementioned studies might be tested experimentally under field conditions by researchers investigating behavior in abiotic and biotic contexts. The latter related programs (Saltzman et al. 2011; Runcie et al. 2013; Weber et al. 2013; Linnen et al. 2013) support the conclusion of Donaldson and Young (2008) that "the molecular basis of social behavior is not beyond the realm of our understanding."

Hayden et al. (2011) portended "the unification of theoretical and experimental work in ecology and neuroscience." Such a "unification" of genetic, physiological, and ecological mechanisms anticipates the synthesis of ecology, neurosciences, and evolutionary biology as well as truly general programs based on highly conserved molecules and switches employed across taxa as "toolkit" functions. Numerous laboratory studies addressing mammalian social behavior and social organization (e.g., "social familiarity," parental care, mate guarding) have been conducted by neuroscientists investigating human "neurogenetics" ("multilevel" *Homo sapiens*) and the related, highly conserved neuropeptides, oxytocin and vasopressin, usually investigated concurrently (rodents, Ferguson et al. (2000); polygynandrous rhesus macaques, *Macaca mulatta,* Chang et al. (2012); Keverne and Curley 2004; cf. Donaldson and Young 2008). Considering "social behavior and social cognition," Insel and Young (2000), found that "certain neuropeptide effects appear to be gender-specific."

9.1   An Integration of Social Neuroscience and Ecology Is on the Horizon

The neurotransmitter, dopamine, is implicated in "social decision making" governing "the anticipation and delivery of rewards" (Foti and Hajcak 2012; Schultz 2006, 2012), including "risky rewards" (Schultz 2012), important responses for mammals in nature. Shultz (2006) reviewed the neurophysiology of "reward and uncertainty," events reliably associated with certain regions of the frontal cortex (rhesus macaques, Chang et al. (2013); humans, Osman (2012)). All of the aforementioned processes pertain to challenges encountered in the heterogeneous conditions associated with mammalian social evolution, and it is noteworthy that a rich program exploring aspects of behavioral ecology and mammalian social neuroscience is ongoing in Michael Platt's laboratory at Duke University (Hayden et al. 2011; Adams et al. 2012; Brent et al. 2013; also see, Selman et al. 2012; Austad and Fischer 1991). Ultimately, the phenomena under discussion will be expressed in energetic terms at each level of biological organization responding to spatiotemporally varying reaction norms (e.g., Dietrich and Horvath 2013). The latter program has the potential to unify understanding of whole-organism regulation of energy (see Dietrich and Horvath 2013; Evans ets al. 2012), expressed as formulae amenable to variations of individuals to their "thermal zones" (see Jones 2012). In addition, the aforementioned research projects pertain to the role that "neural plasticity" plays in opportunistic "decision making," an important aspect of the evolution of flexible phenotypes in the heterogeneous regimes housing mammalian evolution (e.g., Barja 2009). Combined with sociogenomics (see below), a truly integrative comparative ecological "neurogenomics" may be on the horizon (cf. Chandrasekaran et al. 2011).

9.2   Mammalian Sociogenetics: What Genes Do What, and How?

Recently, Hopi Hoekstra's laboratory (Weber et al. 2013; Linnen et al. 2013) has increased our understanding of mammalian genetics with the promise of future applications in sociobiology. Weber et al. (2013) decomposed the behavioral patterns associated with burrow construction in mice (*Peromyscus polionotus, P. maniculatus*). Using a genetic cross design, and consistent with some ethological models of behavior, these authors demonstrated that "complex behaviours" may evolve via "multiple genetic changes each affecting distinct behaviour modules." Rueffler et al. (2012; see Chevin and Lande 2013) discussed the evolution of "functional specialization" (division of labor, modularity) with a general model, arguing that "traits" (behavioral modules, in the present case) can be "mapped" onto performance (e.g., behavior). Following the latter authors' logic, modules associated with "complex burrowing" exhibit "positional effects" in mice since a module's position in a behavioral sequence is expected to have been favored by selection.

Rueffler et al. (2012) also showed, theoretically, that selection will follow modularity "when performance is an accelerating function of the degree of specialization." Thus, consistent with intuition, modularity ("functional specialization") will be favored when its (reproductive) benefits outweigh its costs. The previous model, also, shows that modules working in concert will be favored when the resulting efficiency of an organism is increased. The combined results of Weber et al. (2013) and Rueffler et al. (2012) provide a preliminary template for multilevel evolutionary analyses of the range of behaviors associated with group living and, particularly, with facilitation, including the development of these features. "Touching" (e.g., grooming, sniffing) is a ubiquitous interindividual ("social") motor pattern displayed by many group-living mammals that might serve as a target of research into simple "social" responses often displayed as an element of behavioral sequences. A *caveat* for students of mammalian social evolution is that modular design is expected to trade-off with flexibility (Tabone et al. 2010; Nehring et al. 2012), a notable feature of mammalian repertoires and, possibly, explaining why division of labor is limited in the Class.

The report by Linnen et al. (2013; see Tinbergen 1952) provides a "bridge" from undifferentiated to differentiated phenotypes (Rueffler et al. 2012) by demonstrating a mechanism of "rapid evolution" incorporating "multiple mutations at a single gene." Studying two morphological ecotypes (pelage coloration) of deer mice (*P. maniculatus*), Linnen et al. (2013) mapped "distinct regions within the *Agouti* locus associated with each color trait." Local adaptation, thus, was shown to be "the result of independent selection on many mutations within a single locus, each with a specific effect on an adaptive phenotype. " This process, a type of "rapid evolution," minimizes pleiotropic effects due to increased strength of directional selection for particular traits (see Linnen et al. 2013). Both of the aforementioned papers from Hopi Hoekstra's laboratory have the potential to contribute to our understanding of mammalian phenotypic, including social, evolution since "modularity of biological networks is the key driver of evolvability" (Clune et al. 2013).

The results of Linnen et al. (2013) serve as a template for the evolution of signaler–receiver interactions commonly observed in group-living taxa (e.g., interactions between potential mates, between mother and offspring, between kin). Signals are presumed to have arisen by "rapid evolution," though their ecological and genetic correlates remain relatively unstudied. Female mimicry of male genitalia (e.g., pendulous clitoris in polygynandrous spider monkeys, *Ateles*, this brief Box 8.1; genital hypertrophy, mantled howlers, Jones (1985, 1997)) and similar signals (Jones 1995, 2005) serve as examples of possible research projects for social biologists investigating "rapid evolution" in a mammalian signal and the mechanisms underlying those "functional specializations" ("modules", see Evans et al. (2013)). Such a research program could address questions posed by Weber et al. (2013) and Linnen et al. (2012), as well as provide tests of the general propositions resulting from the work by Rueffler et al. (2012). However, as Hopi Hoekstra pointed out in interviews following the publication of Weber et al. (2013), a remaining challenge for mammalian genetics and genomics is to identify what particular genes do and how they do what they do. These areas of research are in their infancy for students of mammalian social biology (see Table 1 in Robinson et al. 2005).

9.3 The Promise of Mammalian Sociogenomics Has Yet to Be Realized

Studying social insects, Robinson and members of his laboratory (e.g., Whitfield et al. 2003, 2006; Toth et al. 2007; Fischman et al. 2011) analyzed molecular pathways of primitively social and eusocial taxa in order to dissect social evolution. This precise though tedious approach requires significant genomic resources, including knowledge of the effects of genes on phenotypes. These investigators' genomic methods permit within- and between-taxa comparisons; however, knowledge of gene function(s) at the species level is limited for social insects (Fischman et al. 2011) and other groups. Although microarray (gene ontology) analyses do not permit tests of causation, they yield cladograms (Fischman et al. 2011) amenable to quantitative modeling. In addition, knowledge of gene function(s), in particular, the effects of molecular changes, provides information about alternative molecular routes associated with genotype to phenotype pathways and constraints, including ecological ones (Fischman et al. 2011). Whitfield et al. (2006) and Fischman et al. (2011) provided further discussion of the problems encountered with these techniques, including the contingent nature of inferences about specifics of gene action (e.g., epistasis, pleiotropy) and comparative supra-genomic analyses. The issues discussed in these papers should apply, as well, to other synthetic initiatives addressing the analysis of character traits from the genome level, and Robinson and his colleagues (Robinson 2005, 2008) have advanced the "toolkit" paradigm whereby protein diversity is generated by particularly conserved and derived biochemical building blocks.

With mammalian social evolution in mind, Table 3.1 (also see Fig. 3.1) presents preliminary evidence for the idea of an ancient morphological and behavioral "toolkit" in the Class that, along with anatomy and physiology, is expected to correspond to feedback pathways from genes to phenotype to environment and back.

References

Adams GK, Watson KK, Pearson J, Platt ML (2012) Neuroethology of decision-making. Curr Opin Neurobiol 22:982–989

Austad SN, Fischer KA (1991) Mammalian aging, metabolism, and ecology: evidence from the bats and marsupials. J Gerontol 46:B47–B53

Barja I (2009) Decision-making in plant selection during the faecal-marking behaviour of wild wolves. Anim Behav 77:489–493

Brent LJN, Heilbronner SR, Horvath JE, Gonzalez-Martinez J, Ruiz-Lambides A, Robinson AG, Skene JHP, Platt ML (2013) Genetic origins of social networks in Rhesus macaques. Sci Rep 3:1042. doi:10.1038/srep01042

Briga M, Pen I, Wright J (2012) Care for kin: within-group relatedness and allomaternal care are positively correlated and conserved throughout the mammalian phylogeny. Biol Lett 8:533–536

Chandrasekaran S, Ament SA, Eddy JA, Rodriguez-Zas SL, Schatz BR, Price ND, Robinson GE (2011) Behavior-specific changes in transcriptional modules lead to distinct and predictable neurogenomic states. Proc Nat Acad Sci U S A 108:18020–18025

Chang SWC, Gariépy J-F, Platt ML (2013) Neuronal reference frames for social decisions in primate frontal cortex. Nat Neurosci 16:243–250

Chang SWC, Barter JW, Ebitz RB, Watson KK, Platt ML (2012) Inhaled oxytocin amplifies both vicarious reinforcement and self reinforcement in rhesus macaques (*Macaca mulatta*). Proc Nat Acad Sci U S A 109:959–964

Chevin L-M, Lande R (2013) Evolution of discrete phenotypes from continuous norms of reaction. Am Nat 182:13–27

Clune J, Mouret J-B, Lipson H (2013) The evolutionary origins of modularity. Proc Roy Acad Sci B. doi:10.1098/rspb.2012.2863

Dietrich MO, Horvath TL (2013) Hypothalamic control of energy balance: insights into the role of synaptic plasticity. Trends Neurosci 36:65–73

Donaldson ZR, Young LJ (2008) Oxytocin, vasopressin, and the neurogenetics of sociality. Science 322:900–904

Eibl-Eibesfeldt I (1970) Ethology: the biology of behavior. Holt, Rinehart, & Winston, New York

Eibl-Eibesfeldt I (2007) Human ethology. Aldine Transactions (Aldine De Gruyter), Piscataway

Evans AR, Jones D, Boyer AG, Brown JH, Costa DP, Ernest SKM, Fitzgerald EMG, Fortelius M, et al. (2012) The maximum rate of mammalian evolution. Proc Natl Acad Sci U S A 109: 4187–4190

Ferguson JN, Young LJ, Hearn EF, Matzuk MM, Insel TR, Winslow JT (2000) Social amnesia in mice lacking the oxytocin gene. Nat Genet 25:284–288

Fischman BJ, Woodard SH, Robinson GE (2011) Molecular evolutionary analyses of insect societies. Proc Nat Acad Sci U S A 108(Supplement 2):10847–10854.

Foti D, Hajcak G (2012) Genetic variation in dopamine moderates neural response during reward anticipation and delivery: evidence from event-related potentials. Psychophysiology 49: 617–626

Hayden BY, Pearson JM, Platt MI (2011) Neuronal basis of sequential foraging decisions in a patchy environment. Nat Neurosci. doi:10.1038/nn.2856

Insel TR, Young L (2000) Neuropeptides and the evolution of social behavior. Curr Opin Neurobiol 10:784–789

Jones CB (1985) Reproductive patterns in mantled howler monkeys: estrus, mate choice, and copulation. Primates 26:130–142.

Jones CB (1995) Mimicry in primates: implications for heterogeneous conditions. Neotrop Primates 3:69–72

Jones CB (1997) Subspecific differences in vulva size between *Alouatta palliata palliata* and *A. p. mexicana*: implications for assessment of female receptivity. Neotrop Primates 5:46–48

Jones CB (2005) Social parasitism in mammals with particular reference to Neotropical primates. Mastozoológíca Neotrop 12:19–35

Jones CB (2009). The effects of heterogeneous regimes on reproductive skew in eutherian mammals. In: Hager R, Jones CB Reproductive skew in vertebrates: proximate and ultimate causes. Cambridge University Press, Cambridge

Jones CB (2012) Robustness, plasticity, and evolvability in mammals: a thermal niche approach. Springer, New York

Keverne EB, Curley JP (2004) Vasopressin, oxytocin and social behaviour. Curr Opin Neurobiol 14:777–783

Lerner M (1954) Genetic homeostasis. Dover, New York

Lim MM, Wang Z, Olazábal DE, Ren X, Terwilliger EF, Young LJ (2004) Enhanced partner preference in a promiscuous species by manipulating the expression of a single gene. Nature 429:754–757

Linnen CR, Poh Y-P, Peterson BK, Barrett RDH, Larson JG, Jensen JD, Hoekstra HE (2013) Adaptive evolution of multiple traits through multiple mutations at a single gene. Science 339:1312–1316

Nehring V, Boomsma JJ, d'Ettorre P (2012) Wingless virgin queens assume helper roles in *Acromyrmex* leaf-cutting ants. Curr Biol 22:R671–R673

Osman M (2012) The role of reward in dynamic decision-making. Front Neurosci 6. doi:10.3389/fnins.2012.00035

Platt ML (2013) Quote from website. http://neurobiology.duhs.duke.edu/faculty/platt/. Retrieved 26 Oct 2013

Proulx SR (2001) Can behavioural constraints alter the stability of signaling equibrilia? Proc Roy Soc London B 268:2307–2313

Robinson GE, Grozinger CM, Whitfield CW (2005) Sociogenomics: social life in molecular terms. Nat Rev Genet 6:257–270.

Robinson GE, Fernald RD, Clayton DF (2008) Genes and social behavior. Science 322:896–900.

Rueffler C, Hermisson J, Wagner GP (2012) Evolution of functional specialization and division of labor. Proc Nat Acad Sci U S A 109:E326–E335

Runcie DE, Wiedmann RT, Archie EA, Altmann J, Wray GA, Alberts SC, Tung J (2013) Social environment influences the relationship between genotype and gene expression in wild baboons. Phil Trans Roy Soc B. doi:org.10.1098/rstb.2012.0345

Saltzman W, Boettcher CA, Post JL, Abbott DH (2011) Inhibition of maternal behaviour by central infusion of corticotrophin-releasing hormone in marmoset monkeys. J Neuroendocrinol 23:1139–1148

Schultz W (2006) Behavioral theories and the neurophysiology of reward. Ann Rev Psychol 57: 87–115

Schultz W (2012) Risky dopamine. Biol Psychiatry 71:180–181

Selman C, Blount JD, Nussey DH, Speakman JR (2012) Oxidative damage, ageing, and life-history evolution: where now? Trends Ecol Evol 27:570–577

Stowers L, Holy TE, Meister M, Dulac C, Koentges G (2002) Loss of sex discrimination and male-male aggression in mice deficient for TRP2. Science 295:1493–1500

Sutherland WJ, Norris K (2003) Behavioural models of population growth rates: implications for conservation and prediction. In: Sibley RM, Hone J, Clutton-Brock TH Wildlife population growth rates. Cambridge University Press, Cambridge, pp 225–248

Tabone M, Ermentrout B, Doiron B (2010) Balancing organization and flexibility in foraging dynamics. J Theor Biol 266:391–400

Tinbergen N (1952) "Derived" activities: their causation, biological significance, origin, and emancipation during evolution. Quart Rev Biol 27:1–32

Toth AL, Varala K, Newman TC, Miguez FE, Hutchison SK, Willoughby DA, Simons JF, Egholm M, Hunt JH, Hudson ME, Robinson GE (2007) Wasp gene expression supports an evolutionary link between maternal behavior and eusociality. Sciencexpress. www.sciencexpress.org. doi:10.1126/science.1146647, 1–4

Wang H, Duclot F, Liu Y, Wang Z, Kabbaj M (2013) Histone deacetylase inhibitors facilitate partner preference in female prarie voles. Nat Neurosci. doi:10.1038/nn.3420

Weber JN, Peterson BK, Hoekstra HE (2013) Discrete genetic modules are responsible for complex burrow evolution in *Peromyscus* mice. Nature 493:402–405

Whitfield CW, Cziko A-M, Robinson GE (2003) Gene expression profiles in the brain predict behavior in individual honey bees. Science 302:296–299

Whitfield CW, Ben-Shahar Y, Brillet C, Leoncini L, Crauser D, LeConte T, Rodriguez-Zas S, Robinson GE (2006) Genomic dissection of behavioral maturation in the honeybee. Proc Nat Acad Sci U S A 103:16068–16075

Chapter 10
Synopsis

Temperature has long been known to influence metabolism at multiple scales of ecological organization from individuals to ecosystems.

Price et al. (2012)

Abstract This chapter constitutes the present brief's synopsis and coda addressing attempts to formulate general principles of social evolution (synopsis), the possible integration of the latter programs with those treating general laws of energy relations in mammals (synopsis), and priorities for future research in mammalian social biology (coda).

Keywords Convergent evolution · Functional traits · Adaptive traits · Functional diversity · Interspecific trait continuums · General principles

10.1 Synopsis

Numerous research programs are dedicated to the identification and specification of synthetic patterns and principles of social evolution, within and between taxa (Lehmann and Keller 2006). A common assumption of these ventures is that, over ecological and evolutionary time, organisms have responded to similar environmental challenges in similar ways due to fundamental biogeochemical constraints and first principles of ecology (energy dispersion, energy acquisition, consumption, and energy allocation). These assumptions are consistent with the idea that social species have converged on a "similar suite of traits" (Sect. 5.4). However, the disappointing history of searches for "general unifying theories" (GUTs) in Physics and Macroecology raises *caveats* for biologists seeking synthetic formulations of social evolution, not to mention integrating models of social evolution with general principles of energetics (Evans et al. 2013; Hamilton et al. 2011; Evans et al. 1997; see Proulx 1999). Justifying caution, lineage-specific proteins are being identified with increasing frequency, and, some potential mechanisms of social evolution are not well researched (e.g., Table 10.1). Nonetheless, though synthetic approaches are in early stages of investigation, the ideas, methods, data, and models available in

C. B. Jones, *The Evolution of Mammalian Sociality in an Ecological Perspective,* SpringerBriefs in Ecology, DOI 10.1007/978-3-319-03931-2_10,
© Clara B. Jones 2014

the literature demonstrate the utility of proceeding with flexible frameworks enlarging databases, permitting searches for patterns, contingently specifying predictive schemas (Table 10.1). This vision complements that of Keller (1995): "The same conceptual framework can be used to study the social organization of insect and vertebrate societies. Ecological factors, together with internal factors, such as, relatedness, determine the degree of within-group conflict, partitioning of reproduction, and the stable social structure of animals, independent of whether they are ants, birds, or mammals." Expanding the study of unitary integration and sociality, Bourke (2011) has situated the study of social evolution among all transitions to complexity.

10.2   Coda

The synthetic approaches outlined in Table 10.1 address group formation and maintenance, as well as, social traits and their correlates. Synthetic quantitative models showing how sociality *might have* evolved have been attempted. However, we do not know *how* sociality evolved in various regimes inhabited by a variety of conspecific and contraspecific types. Such models require formulations constrained by Hamilton's rule sensitive to competitive contexts, both intra- and inter-group, including responses to these conditions by types varying in their varying trait profiles (Fig. 1.3). The following set of ideas reflect my research interests as well as my opinions about what research topics and approaches require prioritization by students of mammalian social biology. Descriptive studies should collect data appropriate for tests of ideas based on inclusive fitness theory and evolutionary ecology, including those discussed in Chap. 2. The proposed topics are:

1. The formulations of West et al. (2002) suggest that leaving a group may represent altruism ("altruistic dispersal": Taylor et al. 2013) benefiting the relative reproductive success of kin left behind, a model deserving systematic study, including experiments, in association with distance of dispersal (i.e., whether short-distance dispersal creates "viscosity" favoring sociality). It might be expected that low quality dispersers would compete to join groups with the lowest mean l^* values in a population.
2. There is a critical need for systematic studies of "policing," persuasion, coercion, and force within groups of mammals in order to understand tactics and strategies employed by reproductive males and females to manage competition and repress selfishness, particularly where unrelated animals (including humans) co-exist. Mechanisms to manage conflict and social parasitism such as dominance hierarchies and queuing are best viewed in terms of their underlying traits that can be partitioned into causes, and comparative studies of these mechanisms are required in order to understand social evolution in mammals.
3. Comparative studies between termites and social mammals might be instructive since termite workers can be male as well as female (Chap. 5).

Table 10.1 Selected synthetic approaches to social evolution assessed by relevance to group formation (*GF*) and group maintenance (*GM*)

Framework	GF	GM	Comments	References
Qualitative models: pattern–detection strategies (guided "fishing")	?	?	Assembly of large databases of traits, bioinformatics, may capture traits missed by other approaches, inefficient and time consuming	Ongoing at federally funded "cross disciplinary science centers;"
Verbal model I: energetic models	Yes	Yes	Social evolution biased by initial reproductive allocations or energetic investments, based on ecology's "first principles"	Schoener 1971; Trivers 1972
Verbal model II: correlated trait analysis	Yes	Yes	"Trajectories" approach, search for traits systematic, hypothetico-deductive paradigm, "decisions" by type for solitary, singular, or plural breeding, makes predictions about causal factors	Helms Cahan et al. 2002
Verbal model III: Behavioral ecology	Yes	Yes	Emphasis on evolution of mating systems by sexual selection; amenable to quantitative tests, including, experimentation, in context of Ecology's "first principles"	Chapter 6
Multi-level models: spatial mapping approach	?	?	Integrated databases, captures heterogeneous environmental patterns (e.g., climate, resource dispersion), provides no information about dynamics or "patch," amenable to quantitative modeling	Jetz and Rubenstein 2011; this brief Fig. 1.1
Sociogenomics	No	No	Reductionistic, mechanistic, "bottom–up" empirical approach detailing "gene ontologies," identification of conserved biochemical (protein) "toolkits," data correlational, identifies "building blocks" of social evolution, analyses awaiting data from larger base of taxa	Chapter 9
Quasi-theoretical model I: "reproductive skew"	No	Yes	Assumptions have proved problematic, not readily testable (but see Akçay et al. 2012), emphasis on group (reproductive) "productivity"	Vehrencamp 1983, 2000; Emlen 1982, 1995; Frank 1995, 2003; Hager and Jones 2009
Quasi-theoretical model II: biometrical approach to "reproductive skew"	No	Yes	"Animal model" based on quantitative genetics, resolves some problems associated with original skew models, determines heritability of "reproductive share," data required to test model not available	Nonacs and Hager 2011; Nonacs 2010

Despite innovations in verbal and quantitative theory and in empirical approaches, each of these formulations must fulfill requirements of Hamilton's rule, in particular, the differential costs and benefits to related and unrelated types of sociality. Lehmann and Keller (2006) devised a mathematically accessible construct incorporating 35 symbols intended to summarize 38 "influential or original (quantitative) models," finding that each was reducible to Hamilton's rule, including, "group selection" models. Hamilton's rule, likewise, must underlie each model in this table in order to specify how social traits are favored by selection if they are expressed. The parameters of inclusive fitness theory must be explicitly partitioned into variation of discrete causes (abiotic and biotic, including social, stimuli (competition)) and expressed traits (alleles) responding to them. In theory, given a focal individual (FI: see Lehmann and Keller 2006) interacting with a conspecific (a "dyad"), specification of an array of stimuli should predict an array of responses and vice versa, weighted by similarity and dissimilarity of alleles mapped onto interaction rates. Using a functional trait approach permits the researcher to study interindividual interactions independent of conspecific or contraspecific identity and relative to environmental (e.g., patch) traits, often a gradient (e.g., soil moisture, tree size, or yield)

4. Some researchers have pointed out that, in certain conditions, interindividual interactions may appear to be cooperative or altruistic but, upon further inspection, may have resulted from one or another variety of exploitation (e.g., social parasitism: see Otte 1975; Galef 1991; Jones 1986; 1997; 2005; 2007). This apparent dilemma requires systematic investigation, including experiments. Related to the foregoing, because mammalian groups are expected to be "hotspots" for interindividual exploitation, "social parasitism" should be investigated using conventional host-parasite models.

5. Large generalist herbivores have interesting whole-body phenotypes. On the one hand, large body size and long life buffers them from the perturbing effects of fluctuating environments (Selman et al. 2012; but see Marcil-Ferland et al. 2013 for a discussion of costs associated with these traits). On the other hand, physiological, morphological, and behavioral flexibility, as well as large brains, afford relatively rapid, opportunistic, if not "real time," accommodation to changing regimes, particularly, "patch" conditions (see Chap. 9). Such phenotypic robustness (generalized body plan) combined with phenotypic flexibility resulting from anatomies and physiologies sensitive to environmental perturbations (see Thompson et al. 2013) suggest what Ketola (personal communication) terms "the plasticity-environmental canalization continuum," measuring the variability of traits across conditions (e.g., gradients).

6. Within- and between-species comparisons are needed to precisely assess causes and consequences of altruistic or altruistic-like behavior when types demonstrate self-restraint, on the other hand, and when "helping" (donation of some share of a "fitness budget") occurs as a result of persuasion, coercion, force, or exploitation. Where altruistic or altruistic-like behavior is imposed on a type, is the type functioning as an altruist or something else (cheater, the sick or terminal, social parasite ("pay to stay"), desperado, "best of a bad job" strategist, etc.). Such studies might provide empirical, including experimental, clarifications of Hamilton's rule.

The previous suggestions for research are a few of many potential research questions and programs that might advance our understanding of mammalian social evolution. The present brief documents that many mammal groups display dynamic patterns of conflict and cooperation over access to limiting resources, often managed by a variety of "policing" mechanisms (e.g., persuasion, exploitation, coercion, force), as well as sexual segregation. Nevertheless, most mammal groups are, as well, characterized by flexible social architecture, including male tactics and strategies. The latter scenario would seem to suggest that, for the most part, mammalian sociality evolved via the semisocial route (see Chap. 1). However, conflict is expected to arise even among kin because "ego" is related to itself by a coefficient of 1.00. Furthermore, unrelated types may be found in the same groups or networks but display significantly differentiated selection of social partners. In other words, favoring "direct reproduction" may yield the highest reproductive benefits in some conditions, "decisions" yielding benefits from "indirect" reproduction, in others.

Because many mammals are "solitary," they provide an opportunity to study constraints on the evolution of sociality with general import, contributing to the the broad, existing literature based on social insects and social birds. A remarkable feature of Class Mammalia is that females are burdened with an extremely high "reproductive load" (Chap. 8), yet, cooperative breeding and eusociality are relatively uncommon mechanisms employed to minimize energetic costs and to manage competition (see Queller and Strassmann 2010; this brief, Sect. 1.1). Again, this occurrence suggests that social evolution is constrained among mammals, a constraint that may be imposed by the stress and/or unpredictability of heterogeneous regimes increasing interindividual conflicts of interest and rates of dispersal from natal groups (see Jones 2009). On the other hand, and on average, mammalian brains are large relative to body size, possibly correlated with opportunistic decision making in changing environments, conditions that might favor facultative, if not obligate, sociality. Rodents will provide rich tests of hypotheses related to social evolution in mammals, particularly, and in vertebrates, generally. The remarkable variability of sociosexual architectures among hystricognath rodents, including, Caviidae and Bathyergidae, recommends these rodents as models of mammalian social evolution.

References

Akçay E, Meirowitz A, Ramsay KW, Levin SA (2012) Evolution of cooperation and skew under imperfect information. Proc Nat Acad Sci USA 109:14936–14941

Bourke AFG (2011) Principles of social evolution. Oxford University Press, Oxford

Emlen ST (1982) The evolution of helping. I. An ecological constraints model. Am Nat 119:29–39

Evans GB, Brown JH, Enquist BJ (1997) A general model for the origin of allometric scaling laws in biology. Science 276:122–126

Evans JP, van Lieshout E, Gasparini C (2013) Quantitative genetic insights into the coevolutionary dynamics of male and female genitalia. Proc R Soc Lond B doi:.org/10.1098/rspb.2013.0749

Frank SA (1995) Mutual policing and repression of competition in the evolution of cooperative groups. Nature 377:520–522

Frank SA (2003) Repression of competition and the evolution of cooperation. Evolution 57:693–705

Galef BJ (1991) Information centers of Norwegian rats: sites for information exchange and information parasitism. Anim Behav 41:295–301

Hager R, Jones CB (2009) Reproductive skew in vertebrates: proximate and ultimate causes. Cambridge University Press, Cambridge

Hamilton MJ, Davidson AD, Sibly RM, Brown JG (2011) Universal scaling of production rates across mammalian lineages. Proc Roy Soc Lond B 278:560–566

Helms Cahan S, Blumstein DT, Sundström L, Liebig J, Griffin A (2002) Social trajectories and the evolution of social behavior. Oikos 96:206–216

Jetz W, Rubenstein DR (2011) Environmental uncertainty and the global biogeography of cooperative breeding in birds. Curr Biol 21:72–78

Jones CB (1986) Infant transfer behavior in humans: a note on the exploitation of young. Aggress Behav 12:167–173

Jones CB (1997) Social parasitism in the mantled howler monkey, *Alouatta palliata* Gray, (Primates: Cebidae [now Atelidae]): involuntary altruism in a mammal? Sociobiology 30:51–61

Jones CB (2005) Social parasitism in mammals with particular reference to Neotropical primates. Mastozoología Neotrop 12:19–35

Jones CB (2007) The evolution of exploitation in humans: "Surrounded by strangers I thought were my friends." Ethology 113:499–510

Jones CB (2009) The effects of heterogeneous regimes on reproductive skew in eutherian mammals. In: Hager R, Jones CB (eds) Reproductive skew in vertebrates: proximate and ultimate causes. Cambridge University Press, London

Keller L (1995) Social life: the paradox of multiple-queen colonies. Trends Ecol Evol 10:355–360

Lehmann L, Keller L (2006) The evolution of cooperation and altruism—a general framework and a classification of models. J Evol Biol 19:1365–1376

Marcil-Ferland D, Festa-Blanchet M, Martin AM, Pelletier F (2013) Despite catchup, prolonged growth has detrimental fitness consequences in a long-lived vertebrate. Am Nat 182. doi:1086/673534

Nonacs P (2010) Reproductive skew. In: Westneat DE, Fox CW (eds) Evolutionary behavioral ecology. Oxford University Press, Oxford

Nonacs P, Hager R (2011) The past, present and future of reproductive skew theory and experiments. Biol Rev 86:271–298

Otte D (1975) On the role of intraspecific deception. Am Nat 109:239–242

Price CA, Weitz JS, Savage VM, Stegen J, Clarke A, Coomes DA, Dodds PS, Etienne RS, Kerkhoff AJ, McCulloh K, Niklas KJ, Olff H, Swenson G (2012) Testing the metabolic theory of ecology. Ecol Lett 15:1465–1474

Proulx SR (1999) Mating systems and the evolution of niche breadth. Am Nat 154:89–98

Queller DC, Strassmann JE (2010) Evolution of complex societies. In: Westneat DE, Fox CW (eds) Evolutionary behavioral ecology. Oxford University Press, Oxford

Schoener TW (1971) Theory of feeding strategies. Ann Rev Ecol Syst 2:369–404

Selman C, Blount JD, Nussey, Speakman JR (2012) Oxidative damage, ageing, and life-history evolution: where now? Trends Ecol Evol 27:570–577

Taylor TB, Rodrigues AMM, Gardner A, Buckling A (2013) The social evolution of dispersal with public goods cooperation. J Evol Biol. doi:10:1111/jeb.12259

Thompson GJ, Hurd PL, Crespi BJ (2013) Genes underlying altruism. Biol Lett 9. doi:10.1098/rsbl.2013.0395

Trivers RL (1972) Parental investment and sexual selection. In: Campbell B (ed) Sexual selection and the descent of man, 1871–1971. Aldine, New York

Vehrencamp SL (1983) A model for the evolution of despotic versus egalitarian societies. Anim Behav 31:667–682

Vehrencamp SL (2000) Evolutionary routes to joint-female nesting in birds. Behav Ecol 11:334–344

West SA, Pen I, Griffin AS (2002) Cooperation and competition between relatives. Science 296: 72–75

Index

C. B. Jones, *The Evolution of Mammalian Sociality in an Ecological Perspective,*
SpringerBriefs in Ecology, DOI 10.1007/978-3-319-03931-2,
© Clara B. Jones 2014